PMP® Exam Practice Test and Study Guide

Tenth Edition

Dr. Ginger Levin, PMP, PgMP
J. LeRoy Ward, PMP, PgMP, PfMP, CSM

CRC Press is an imprint of the
Taylor & Francis Group, an **informa** business
AN AUERBACH BOOK

CRC Press
Taylor & Francis Group
6000 Broken Sound Parkway NW, Suite 300
Boca Raton, FL 33487-2742

CRC Press is an imprint of Taylor & Francis Group, an Informa business

Printed on acid-free paper
Version Date: 20150810

International Standard Book Number-13: 978-1-4987-5282-4 (Hardback)

Visit the Taylor & Francis Web site at
http://www.taylorandfrancis.com

and the CRC Press Web site at
http://www.crcpress.com

Contents

Preface vii
Acknowledgements xi
Introduction xiii
Authors xvii
Acronyms xix
Project Integration Management 1
Project Scope Management 31
Project Time Management 61
Project Cost Management 93
Project Quality Management 123
Project Human Resource Management 153
Project Communications Management 185
Project Risk Management 213
Project Procurement Management 245
Project Stakeholder Management 271
An Overview of the Domains 299
Practice Test 309
Answer Sheet 357
Appendix: Study Matrix 363
Answer Key 373
References 439

Preface

We have been helping people to prepare for the project management professional (PMP®)* certification exam since the *A Guide to the Project Management Body of Knowledge* PMBOK® Guide was published by the Project Management Institute (PMI®) in 1996. We then prepared the first edition of this Study Guide.

In 1996, there were approximately 1,000 people who were certified as Project Management Professional (PMP®), while in 2015, there are approximately 700,000 PMP®s around the world. Throughout the years, we have updated our PMP Practice Test and Study Guide books so that now this book is in its 10th Edition.

Since our first Study Guide book was published, it has become quite clear that most prospective exam takers (ourselves included when we studied for the exam) ask two questions when they decide to earn PMP® certification: "What topics are covered on the exam?" and "What are the questions like?" Not surprisingly, some of the most sought-after study aids are practice tests, which are helpful in two ways: first, taking practice tests increases your knowledge of the kinds of questions, phrases, terminology, and sentence construction that you will encounter on the "real" exam; and second, taking practice tests provides an opportunity for highly concentrated study by exposing you to a breadth of project management content generally not found in a single reference source.

We initiated and continued the development of this specialty publication with only one simple goal in mind: to help you study for, and pass, the PMP® certification exam. Because (PMI®)† does not sell "past" exams to prospective certification candidates for study purposes, the best anyone can do is to develop practice test questions that are as representative of the real questions as possible. And that is exactly what we have done. This tenth edition contains 600 practice test questions.

As we developed this publication, we have worked hard to make the questions difficult ones yet representative of what you may encounter on the actual exam. Having attained the PMP®, we know it is a difficult exam, and one that requires study and dedicated effort.

---

* "PMP" is a certification mark of the Project Management Institute, Inc., which is registered in the United States and other nations.

† "PMI" is a service and trademark of the Project Management Institute, Inc., which is registered in the United States and other nations.

The result of our effort is the *PMP® Exam: Practice Test and Study Guide.* This tenth edition contains study hints, a list of exam topics, and 40 multiple-choice questions for each of the ten knowledge areas presented in *A Guide to the Project Management Body of Knowledge Fifth Edition (PMBOK® Guide).**

And, as in previous editions, this edition includes a plainly written rationale for each correct answer, with a reference to the appropriate pages in the *PMBOK® Guide*, and in some cases to the other sources.

We have listened to your comments and have expanded our rationales to discuss in more depth why an answer is correct, often by explaining why the other answers are incorrect, and to help you broaden your understanding of the concepts in the answer. We also have provided some study tips in these rationales.

You additionally will find a reference to one or more of the five process groups, Initiating, Planning, Executing, Monitoring and Controlling, or Closing at the end of each rationale. Those references are important because they give you an understanding of the types of exam questions that fall within each of these five major project management performance domains.

What is different, however, with this tenth edition is each question also contains a reference to the June 2015, "*Examination Content Outline* (ECO)", published by PMI, which is the blueprint for the exam. You can download the ECO from the PMI web site, and you will see it consists of numerous tasks associated with the domains on the exam. These domains are the process groups, and we have a chapter in this tenth edition that describes each domain in more depth. This tenth edition is completely aligned with this new ECO.

This tenth edition includes scenario-based questions, which may comprise approximately 50 percent of the questions found on the PMP® exam. It omits many of the purely definitional questions, but focuses on the key concepts you need to understand and study. Additionally, we have reduced the length of many of our questions, to more accurately represent the real exam.

We have included a completely original, 200-question on-line practice test, which is also in the book. We have a Study Matrix in this edition as well. The matrix is included as an appendix. The matrix, which is based on PMI®'s *Role Delineation Study,* the basic document for the *ECO,* will help you to use the 200-question exam to its full advantage. It provides a way for you to assess your strengths and weaknesses in each performance domain and to identify areas that require further study.

A special note to those who speak English as a second language (ESL): Our experience in teaching project management programs around the world has shown that most of our ESL clients understand English well enough to pass the PMP® exam *as long as they know the content.* Nevertheless, in an effort to avoid adding to your frustration before taking the exam, we have painstakingly reviewed each question and answer in the Study Guide in the knowledge areas

---

* "PMBOK" is a trademark of the Project Management Institute, Inc., which is registered in the United States and other nations.

and in the practice test to ensure that we did not use words, terms, or phrases that could be confusing to those who are not fluent in English.

Although the language issue may concern you, and rightfully so, the only difference between you and those who speak English as their first language is the amount of time it takes to complete the exam. We know of only one person who did not have enough time, and that individual was able to complete all but two questions. We would suggest, therefore, that if you can grasp the content expressed in this publication, a few colloquialisms or ambiguous terms on the real exam will not ultimately determine whether you pass or fail: Your subject matter knowledge will do that!

Earning the PMP® certification is a prestigious accomplishment. But studying for it need not be difficult if you use the tools available. You may want to include our companion piece, *PMP® Exam Challenge! Sixth Edition* in your study plan if you have not already done so. In an easy flash-card format, it too provides many opportunities to become thoroughly familiar and comfortable with the project management body of knowledge.

On-line Practice Test

To access this practice test, please contact pmptest10@ittoday.info. The on-line test lasts four hours and has a clock that ticks down to show you how much time you have left in the four hours for the exam. Once you have answered all the questions or time runs out, you will be scored by domain and also through a proprietary algorithm we have prepared that shows based on our questions, you will see whether you were Proficient, Moderately, or Below in a domain.

Good luck on achieving your PMP® certification!

Ginger Levin
Lighthouse Point, Florida

J. LeRoy Ward
Madison, Connecticut

Acknowledgements

Our special thanks to our editor, John Wyzalek, who personally edited each question and then proofread the entire book later under a tight deadline; our publisher, Rich O'Hanley, who sets the stage for success at CRC; our production team including Randy Burling, who gave this book a special priority to make sure it would be available to you as the exam changed; and Jessica Vakili, who worked under a very tight schedule; cover designer Elise Weinger; and Michelle Rivera-Spann and her team in marketing.

Introduction

The PMP® exam contains 200 questions, of which 25 questions will not be included in the pass/fail determination. These "pretest" items, as PMI® calls them, will be randomly placed throughout the exam to gather statistical information on their performance to determine their use for future exams. The questions on the exam are distributed as follows by process group following the *PMI PMP Examination Content Outline (ECO)—June 2015.*

- 13% or 26 questions relate to Initiating the Project
- 24% or 48 questions relate to Planning the Project
- 31% or 62 questions relate to Executing the Project
- 25% or 50 questions relate to Monitoring and Controlling the Project
- 7% or 14 questions relate to Closing the Project

For the practice test in this book, we provide all 200 questions as if they were real questions, and the percentages above are applied to the 200 questions. There are no pretest questions in our practice exam. We also have linked each question to a specific task in one of the domains in the ECO especially since some of the questions on the PMP exam are based on the tasks in the ECO, and some tasks are not covered in *A Guide to the Project Management Body of Knowledge (PMBOK® Guide)* Fifth Edition.

There also are not a certain number of the scored 175 questions that you must answer correctly. PMI® explains in its *PMP Credential Handbook*—revised 26 June 2015, it generates a pass/fail score based on proficiency levels in the answers in each of the five process groups or domains. This approach means that you will learn whether you were Proficient, Moderately Proficient or Below Proficient in each process group.

PMI® defines on page 34 of the Handbook these terms as follows:

- Proficient—indicates performance is above the average level of knowledge in this chapter. (Note: We interpret chapter to mean domain)
- Moderately Proficient—performance is at the average level of knowledge in this chapter
- Below Proficient—performance is below average of knowledge in this chapter

These scores are established based on a psychometric analysis as PMI® uses subject matter experts throughout the world to establish a point in which it feels a candidate for the exam should be able to pass a question and establishes a level of difficulty for each question. We do not know the algorithm PMI® uses to establish the proficiency levels. Our practice exam, therefore, scores each question equally, which means you will learn the number of questions you answer correctly in each of the five process groups. Obviously, your goal is to be Proficient, and our goal is that this book can assist you in your quest to attain the Proficient level.

To use the study guide effectively, work on one section at a time. It does not matter which one you choose first. Start by reading the study areas. They provide useful background on the content of the PMP® exam and identify the emphasis placed on various topics. Familiarize yourself with the major topics listed. Then answer the 40 practice questions, recording your answers on the sheet provided. Finally, compare your answers with those in the answer key. The rationales provided should clarify any misconceptions you may have had, and the process group or domain designations will give you an understanding of the types of questions you might see on the exam that relate to those process groups. We have expanded in this tenth edition the rationales for the correct answer to assist you in understanding why the answer is correct and provided more details to help in greater understanding on the topic. For further study and clarification, you may want to consult the bibliographic reference.

After you have finished answering the questions that follow each section, it is time to take the completely rewritten and original, 200-question practice test that has the number of questions by domain or process group in the ECO. Note your answers on the sheet provided, compare your answers to the answer key, and use the Study Matrix in the Appendix to determine what areas you need to study further.

We also have this test on-line. To access it, please contact pmptest10@ittoday.info. The on-line test lasts four hours and has a clock that ticks down and shows you how much time you have left in the four hours for the exam. At the end, you will see your score by domain and also through a proprietary algorithm we have prepared that shows based on our questions whether you were Proficient, Moderately Proficient, or Below Proficient in a domain.

It is our recommendation that you take this four-hour on-line test first to see your areas of strength and areas of improvement after you feel you have mastered the material and are ready to take the PMP. Then, try it again. You then have a benchmark and can see if you do need further study.

To make the most of this book, use it regularly. Take and retake the practice test in the book. Photocopy the answer sheet in order to have a clean one each time you retake the test. Our suggestion is to score at least 80% of the questions in our practice test and on-line test correctly before you take the PMP® exam.

You may want to convene a study group to compare your answers with those of your colleagues. This method of study is a powerful one. You will learn more from your colleagues than you ever thought possible! Make sure you have a solid understanding of the exam topics that are provided in each section. Consult our extensive bibliography, or other sources you have found useful, for further independent study. And, most important, create a study plan and stick to it. Your chances of success are raised dramatically when you dedicate yourself to your goal.

We also suggest that when you go to the testing center, you go with a completely positive attitude. By this time, you have done your very best to master the project management concepts you will see on the exam.

Authors

Dr. Ginger Levin is a senior consultant, author, and educator in the project management field with 49 years of experience. Her specialty areas are portfolio management, program management, change management, the project management office, metrics, and maturity assessments. She is certified as a PMP® and PgMP®. She was the second person in the world to receive the PgMP. In 2014, she won PMI's Eric Jenett Award.

In addition, Dr. Levin is an Adjunct Professor for the University of Wisconsin-Platteville where she teaches in its M.S. in Project Management Program, and for SKEMA University, France, in its doctoral programs in project management. Six of her students have won the Project Management Institute's (PMI®) Student Paper of the Year Award.

In consulting, she has served as project manager in numerous efforts for *Fortune* 500 and public sector clients, including Genentech, Cargill, Abbott Vascular, UPS, Citibank, the Food and Drug Administration, General Electric, SAP, EADS, John Deere, Schreiber Foods, TRW, New York City Transit Authority, the U.S. Joint Forces Command, and the U.S. Department of Agriculture. Prior to her work in consulting, she held positions of increasing responsibility with the U.S. Government, including the Federal Aviation Administration, Office of Personnel Management, and the General Accounting Office.

Dr. Levin is the editor, author, and co-author of approximately 20 books and is a series editor on advanced books in portfolio and program management for CRC Press. She presents regularly to the Project Management Institute and other organizations.

Dr. Levin received her doctorate in Information Systems Technology and Public Administration from The George Washington University and received the *Outstanding Dissertation Award* for her research on large organizations.

In her spare time, she is an active volunteer for PMI.

Please see: ***linkedin.com/in/gingerlevin***

J. LeRoy Ward is President of Ward Associates, offering project, program and portfolio consulting and advisory services to corporations worldwide. Formerly, Mr. Ward was the Executive Vice President of ESI International responsible for R&D, Product Strategy and Consulting. Additionally, he spent more than 16 years with four U.S. federal agencies in the fields of cartography and information technology.

Mr. Ward has authored nine publications and more than 40 articles in project, program and portfolio management, which have appeared in such publications as the *Financial Times, PM Network, ProjectManager Today, CIO*, and other notable trade and business publications. He authored the *Dictionary of Project Management Terms (3rd ed.)*; with Ginger Levin, he authored the *PMP® Exam Practice Test and Study Guide, PMP® Exam Challenge, PgMP® Exam Practice Test and Study Guide, PgMP® Exam Challenge* and, *Program Management Complexity: A Competency Model*; and with Carl Pritchard, *The Portable PMP® Prep: Conversations on Passing the PMP® Exam.*

A dynamic and popular speaker and keynote presenter, Mr. Ward frequently speaks on project management and related topics at professional associations, client events, and conferences worldwide. Since 1991, he has worked to provide the most comprehensive guidance to project management practitioners, helping them pass the PMP Exam.

Mr. Ward holds Bachelor of Science and Master of Science degrees from Southern Connecticut State University and a Master of Science degree in technology management from The American University where he was inducted into Phi Alpha Alpha, the National Honorary Society for Public Affairs and Administration. He is an alumnus of the U.S. General Services Administration's *Trail Boss* Program (for major systems acquisitions) and the Federal Executive Institute, the U.S. Government's premiere leadership institute.

He also holds the PMP, PgMP, and PfMP® certifications from the Project Management Institute (PMI®); the CSM (Certified Scrum Master) credential from the Scrum Alliance; and, is certified by both The George Washington University School of Business, and the Stanford University School of Engineering, as a "certified project manager." Mr. Ward is also the recipient of the prestigious 2013 Eric Jenett Project Management Excellence Award from PMI.

Please see: ***linkedin.com/in/jleroyward***

Dr. Levin and Mr. Ward have been helping people prepare for the PMP® since 1996, when the first edition of this Practice Test and Study Guide and the *PMP Challenge!* were published. By attending and presenting at PMI Congresses and to PMI Chapters around the world and reading as much as possible in the field, they keep up with the best practices to follow. While they each have other publications on their own, they have collaborated on a comparable Practice Test and Study Guide for the Program Management Professional (PgMP®) exam and a *PgMP® Challenge*. They authored *Program Management Complexity: A Competency Model*, published in 2011, and led the development of ESI's Maturity Models: *PortfolioFramework™, Program Framework™* and *ProjectFramework™*. Further, they are series editors for CRC in books on project and program management.

Acronyms

AC	actual cost
BAC	budget at completion
CAPM®	Certified Associate in Project Management
CCB	Change Control Board
CEO	chief executive officer
CPI	cost performance index
CPM	critical path method
CSM	Certified Scrum Master
CV	cost variance
EAC	estimate at completion
ECO	Examination Content Outline
EMV	expected monetary value
ERP	enterprise resource planning
ESL	English as a second language
ETC	estimate to complete
EV	earned value
EVA	economic value added
EVM	earned value management
ID	identification
IFB	invitation for bid
ISO	International Organization for Standardization
IT	information technology
LCC	life-cycle cost
MRP	material requirements planning
OBS	organizational breakdown structure
PDM	precedence diagramming method
PERT	program evaluation and review technique
PfMP®	Portfolio Management Professional
PgMP®	Program Management Professional
PMP	performance measurement baseline
PMBOK® Guide	*A Guide to the Project Management Body of Knowledge*
PMI®	Project Management Institute
PMIS	project management information system

PMM	Project Management Methodology
PMO	program management office
PMP®	Project Management Professional
PV	planned value
QA	quality assurance
RACI	responsible, accountable, consult, inform
RAM	responsibility assignment matrix
RBS	resource breakdown structure
RBS	risk breakdown structure
R&D	research and development
RF	radio frequency
RFP	request for proposal
ROI	return on investment
SD	standard deviation
SIPOC	suppliers, inputs, process, outputs, customers
SPC	statistical process control
SPI	schedule performance index
SV	schedule variance
SWOT	strengths-weaknesses-opportunities-threats
TCPI	to-complete performance index
VAC	variance at completion
WBS	work breakdown structure

Project Integration Management

Study Hints

The Project Integration Management questions on the PMP® certification exam address critical project management functions that ensure coordination of the various elements of the project. As the *PMBOK® Guide* explains the processes focus on integration activities designed to ensure project success; therefore, integration characteristics involve unification, consolidation, communication, and integrative activities. Project Integration Management involves making decisions about resource use, trade-offs among competing objectives and alternatives, and managing the interdependencies between the ten knowledge areas. It addresses project initiation with developing a project charter, developing a project plan, directing and managing the project work, monitoring and controlling the project work, performing integrated change control, and closing the project or a phase. These six processes not only interact with one another but also interact with processes in the other nine knowledge areas. It is important to note PMI®'s view that integration occurs in other areas as well. For example, project scope and product scope need to be integrated, project work needs to be integrated with other ongoing work of the organization, and deliverables from various technical specialties need integration.

The Project Integration Management questions are relatively straightforward. Most people find them to be fairly easy. But because they cover so much material, including all five process groups as well as some of the materials from the first two chapters in the *PMBOK® Guide*, you do need to study them carefully to become familiar with PMI®'s terminology and perspectives. *PMBOK® Guide* Figure 4-1 provides an overview of the structure of Project Integration Management. Know this chart thoroughly.

Following is a list of the major Project Integration Management topics. Use it to help focus your study efforts on the areas most likely to appear on the exam.

Major Topics

Project, program, and portfolio definitions
Project management definition
Project life cycle
Project management office (PMO)
Project process groups

- Initiating
- Planning
- Executing
- Monitoring and Controlling
- Closing

Business case
Project statement of work
Develop project charter

Enterprise environmental factors
Organizational process assets
Project management information system (PMIS)
Facilitation techniques
Analytical techniques
Expert judgment
Project management plan

Direct and manage project work
Key management reviews
Corrective and preventive action
Deliverables
Work performance information
Meetings
Project baselines
Subsidiary plans
Standards and regulations

Monitor and control project work
Validated changes
Forecasts

Integrated change control
Change requests
Change control meetings
Change control procedures

Change management plan
Configuration management plan
Change log
Approved change requests

Close project or phase
Administrative closure procedure
Accepted deliverables
Product, service, or result transition
Lessons learned

Practice Questions

INSTRUCTIONS: Note the most suitable answer for each multiple-choice question in the appropriate space on the answer sheet.

1. You work for a software development company that has followed the waterfall development model for more than 20 years. Lately, a number of customers have complained that your company is taking too long to complete its projects. You attended a class on agile development methods and believe that if the company used the agile approach, it could provide products to clients in a shorter time period. However, it would be a major culture change to switch from the waterfall methodology to the agile approach and to train staff members in this new approach. You mentioned this idea to the director of the PMO, and although she liked the idea, she would need approval from the company's portfolio review board to move forward with it. She suggested that you document this idea in a—
   a. Business need description
   b. Product scope description
   c. Project charter
   d. Business case

2. You are managing a large project with 20 key internal stakeholders, eight contractors, and six team leaders. You must devote attention to effective integrated change control. This means you are concerned primarily with—
   a. Reviewing, approving, and controlling changes
   b. Maintaining baseline integrity, integrating product and project scope, and coordinating change across knowledge areas
   c. Integrating deliverables from different functional specialties on the project
   d. Establishing a change control board that oversees the overall project changes

3. You plan to hold a series of meeting as you execute the project plan. While different attendees will attend each meeting, a best practice to follow is to:
   a. Group stakeholders into categories to determine which ones should attend each meeting
   b. Not mix the types of meetings on your project
   c. Be sensitive to the fact that stakeholders often have very different objectives and invite them to determine the meeting's agenda
   d. Recognize that roles and responsibilities may overlap so focus on holding meetings primarily for decision making

4. You are the project manager in charge of developing a new shipping container for Globus Ocean Transport, which needs to withstand winds of 90 knots and swells of 30 meters. In determining the dimension of the container and the materials to be used in its fabrication, you convene a group of knowledgeable professionals to gather initial requirements, which will be included in the—

   a. Project charter
   b. Bill of materials
   c. WBS
   d. Project Statement of Work

5. You have assembled a core team to develop the project management plan for the next generation of fatigue fighting drugs. The science is complex, and the extended team of researchers, clinicians, and patients for trials exceeds 500 people. The content of your project management plan will be directed primarily by two factors. They are—

   a. Project complexity and the capability of resources
   b. Number of resources and project schedule
   c. Team member experience and budget
   d. Application area and complexity

6. When you established the change control board for your avionics project, you established specific procedures to govern its operation. The procedures require all approved changes to baselines to be reflected in the—

   a. Performance measurement baseline
   b. Change management plan
   c. Quality assurance plan
   d. Project management plan

7. You are beginning a new project staffed with a virtual team located across five countries. To help avoid conflict in work priorities among your team members and their functional managers, you ask the project sponsor to prepare a—

   a. Memo to team members informing them that they work for you now
   b. Project charter
   c. Memo to the functional managers informing them that you have authority to direct their employees
   d. Human resource management plan

8. The purpose of economic value added (EVA) is to—

   a. Determine the opportunity costs associated with the project
   b. Determine a non–time-dependent measure of profit or return
   c. Assess the net operating profit after taxes
   d. Evaluate the return on capital percent versus the cost of capital percent

9. Facilitation techniques are used throughout project management. Your company is embarking on a project to completely eliminate defects in its products. You are the project manager for this project, and you are developing your project charter. To assist you, which of the following facilitation techniques did you use?

   a. Surveys
   b. Delphi approach → Business forecasting
   c. Meeting management
   d. Focus groups

10. The direct and manage project work process truly is important in project management. It affects many other key processes and uses inputs from others. Working with your team at its kickoff meeting, you explain the key benefit of this process is to—

    a. Implement approved changes
    b. Provide overall management of the project work
    c. Lead and perform activities in the project management plan
    d. Perform activities to accomplish project objectives

11. You are managing a project in an organization that is characterized by rigid rules and policies and strict supervisory controls. Your project, sponsored by your CEO who is new to the company, is to make the organization less bureaucratic and more participative. You are developing your project management plan. Given the organization as it now is set up, as you prepare your plan, you can use which of the following organizational process assets—

    a. Guidelines and criteria
    b. Project management body of knowledge for your industry
    c. Organizational structure and culture
    d. The existing infrastructure

12. You are fairly new to managing a project but have been a team member for many years. You are pleased you were selected to manage your company's 2018 model line of hybrid vehicles. You are now planning your project and have been preparing the subsidiary plans as well. You realize some project documents also are required to help manage your project. An example of one that you believe will be especial helpful is the—

    a. Business case
    b. Key performance indicators
    c. Project management information system
    d. Project statement of work

13. You work for a telecommunications company, and when developing a project management plan for a new project, you found that you must tailor some company processes because the product is so different than those products typically produced by your company. To tailor these processes, you will follow—
    a. Standardized guidelines and work instructions
    b. Stakeholder risk tolerances
    c. Expert judgment
    d. Structure of your company

14. You are implementing a project management methodology for your company that requires you to establish a change control board. Which one of the following statements best describes a change control board?
    a. Recommended for use on all (large and small) projects
    b. Used to review, evaluate, approve, delay, or reject changes to the project
    c. Managed by the project manager, who also serves as its secretary
    d. Composed of key project team members

15. An automated tool, project records, performance indicators, data bases, and financials are examples of items in—
    a. Organizational process assets
    b. Project management information systems
    c. Project management planning approaches
    d. The tools and techniques for project plan development

16. You realize that projects represent change, and on your projects, you always seem to have a number of change requests to consider. In your current project to manage the safety of the nation's cheese products and the testing methods used, you decided to prepare a formal change management plan. An often overlooked type of change request is—
    a. Adding new subject matter experts to your team
    b. Updates
    c. Work performance information
    d. Enhancing the reviews performed by your project's governance board

17. You have been directed to establish a change control system for your company, but must convince your colleagues to use it. To be effective, the change control system must include—
    a. Procedures that define how project documents may be changed
    b. Specific change requests expected on the project and plans to respond to each one
    c. Performance reports that forecast project changes
    d. A description of the functional and physical characteristics of an item or system

18. You are working on the next generation of software for mobile phones for your telecommunications company. While time to market is critical, you know from your work on other projects that management reviews can be helpful and plan to use them on your project. You are documenting them as part of your—

    a. Governance plan
    b. Change management plan
    c. Performance reviews
    d. Project management plan

19. Your cost control specialist has developed a budget plan for your project to add a second surgical center to the Children's Hospital. As you analyze cash flow requirements, you notice that cash flow activity is greatest in the closing phase. You find this unusual because on most projects the largest portion of the budget is spent—

    a. Initiating
    b. Monitoring and Controlling
    c. Controlling
    d. Executing

20. You are project manager for a systems integration effort and need to procure the hardware components from external sources. Your subcontracts administrator has told you to prepare a product description, which is referenced in a—

    a. Project statement of work
    b. Contract scope statement
    c. Request for proposal
    d. Contract

21. Because your project is slated to last five years, you believe rolling wave planning is appropriate. It provides information about the work to be done—

    a. Throughout all project phases
    b. For successful completion of the current project phase
    c. For successful completion of the current and subsequent project phases
    d. In the next project phase

22. You want to minimize the impact of changes on your project, yet you want to ensure that change is managed when and if it occurs. This can be done through each of the following ways EXCEPT—

    a. Rejecting requested changes
    b. Approving changes and incorporating them into a revised baseline
    c. Documenting the complete impact of requested changes
    d. Ensuring that project scope changes are reflected in changes to product scope

23. You are managing a project to introduce a new product to the marketplace that is expected to have a very long life. In this situation, the concept of being *temporary,* which is part of the definition of a project,—

    a. Does not apply because the project will have a lasting result
    b. Does not apply to the product to be created
    c. Recognizes that the project team will outlive the actual project
    d. Does not apply because the project will not be short in duration

24. When closing a project, it is a best practice to—

    a. Update the project documents
    b. Prepare a sustainment plan for the project's benefits
    c. Measure product scope against the project management plan
    d. Review the scope baseline

25. All the following are project baselines that are generally part of the project management plan EXCEPT—

    a. Technical
    b. Scope
    c. Time
    d. Cost

26. You are responsible for a project management training curriculum that is offered throughout the organization. In this situation, your intangible deliverables are—

    a. Employees who can apply the training effectively
    b. Training materials for each course
    c. Certificates of completion for everyone who completes the program
    d. The training curriculum as advertised in your catalog

27. Working on your project management training curricula project, you decided it would be beneficial to you to become an active member of the Project Management Institute as part of the objectives of your project is to ensure it is aligned with PMI®'s best practices. To complement PMI®'s *Work Breakdown Structure Practice Standard,* you learned PMI® was requesting volunteers to participate in development of a similar standard on the Scope Statement. You volunteered, and now the Standard is issued. This is an example of—

    a. Improving your own competency as a project manager
    b. Corrective action
    c. Preventive action
    d. A requirement for you to immediately update your project management plan

28. Ideally, a project manager should be selected and assigned at which point in the project life cycle?

    a. During the initiating processes
    b. During the project planning process
    c. At the end of the concept phase of the project life cycle
    d. Prior to the beginning of the development phase of the project life cycle

29. Closing a project phase should not be delayed until project completion because—

    a. Useful information may be lost
    b. The project manager may be reassigned
    c. Project team members may be reassigned by that time
    d. Sellers are anxious for payments

30. As you are working on your telecommunications project, even though you are using agile methods, you realize you are preparing an extensive amount of data and information. You regularly share data with your project team. Your last team meeting focused on the number of change requests and also the start and finish dates of activities in your schedule. They are examples of—

    a. Key performance indicators
    b. Work performance reports
    c. Work performance data
    d. Work performance information

31. Project management processes describe project work, while product-oriented management processes specify the project's product. Therefore, a project management process and a product-oriented management process—

   a. Overlap and interact throughout the project
   b. Are defined by the project life cycle
   c. Are concerned with describing and organizing project work
   d. Are similar for each application area

32. The close project or phase process addresses actions and activities concerning all of the following EXCEPT—

   a. Completion or exit criteria for the project or phase have been met
   b. Stakeholder approval that the project has met their requirements
   c. Review of the project and/or phase information for potential future use
   d. Documentation that completed deliverables have been accepted

33. You are a personnel management specialist recently assigned to a project team working on a team-based reward and recognition system. The other team members also work in the human resources department. The project charter should be issued by—

   a. The project manager
   b. The client
   c. A sponsor
   d. A member of the PMO who has jurisdiction over human resources

34. Your project is proceeding according to schedule. You have just learned that a new regulatory requirement will cause a change in one of the project's performance specifications. To ensure that this change is incorporated into the project management plan, you should—

   a. Call a meeting of the change control board
   b. Change the WBS, project schedule, and project plan to reflect the new requirement
   c. Prepare a change request
   d. Immediately inform all affected stakeholders of the new approach to take on the project

35. Different types of project phases are used on projects, and each phase culminates in the completion of at least one deliverable. The high-level nature of these phases means they are an element of the project life cycle. Some phases start before others complete. If this approach is followed, it may result in—

    a. An increase in the number of issues
    b. Increasing the schedule
    c. The need for a CCB
    d. More rework

36. Assume your company is a leader in the market in production of cereal products. It has been in this market for over 50 years. You are the project manager for a new product that is a derivative from the company's core product. As you determine a life cycle for this project, you believe you should follow one that is—

    a. Incremental
    b. Predictive
    c. Iterative
    d. Adaptive

37. Oftentimes when a project is terminated, senior managers will replace the project manager with an individual who is skilled in closing out projects. If this is done, the first step for the termination manager should be to—

    a. Notify all relevant stakeholders of the termination
    b. Complete the lessons learned report
    c. Conduct an immediate review of the work packages
    d. Review the status of all contracts

38. On your project you want to avoid bureaucracy, so you adopt an informal approach to change control. The main problem with this approach is—

    a. There is no "paper trail" of change activity
    b. Regular disagreements between the project manager and the functional manager will occur
    c. There are misunderstandings regarding what was agreed upon by stakeholders
    d. There is a lack of sound cost estimating to assess the change's impact

39. Projects are supposed to succeed, not fail. However, termination is an option to consider when all but which one of the following conditions exist?

    a. The customer's strategy has changed.
    b. There are new stakeholders.
    c. Competition may make the project results obsolete.
    d. The original purposes for the project have changed.

40. All projects involve some extent of change, because they involve work that is unique in some fashion. Therefore, it is important that a project management plan includes a—

    a. Description of the change request process
    b. Configuration management plan
    c. Methodology for preventive action to avoid the need for excessive changes
    d. A work authorization system

Answer Sheet

1.	a	b	c	d	21.	a	b	c	d
2.	a	b	c	d	22.	a	b	c	d
3.	a	b	c	d	23.	a	b	c	d
4.	a	b	c	d	24.	a	b	c	d
5.	a	b	c	d	25.	a	b	c	d
6.	a	b	c	d	26.	a	b	c	d
7.	a	b	c	d	27.	a	b	c	d
8.	a	b	c	d	28.	a	b	c	d
9.	a	b	c	d	29.	a	b	c	d
10.	a	b	c	d	30.	a	b	c	d
11.	a	b	c	d	31.	a	b	c	d
12.	a	b	c	d	32.	a	b	c	d
13.	a	b	c	d	33.	a	b	c	d
14.	a	b	c	d	34.	a	b	c	d
15.	a	b	c	d	35.	a	b	c	d
16.	a	b	c	d	36.	a	b	c	d
17.	a	b	c	d	37.	a	b	c	d
18.	a	b	c	d	38.	a	b	c	d
19.	a	b	c	d	39.	a	b	c	d
20.	a	b	c	d	40.	a	b	c	d

Answer Key

1. d. Business case

   The business case is used to provide the necessary information to determine whether or not a project is worth its investment. It is used to justify the project and typically contains a cost-benefit analysis and a business need. [Initiating]

   PMI®, *PMBOK® Guide*, 2013, 69
   PMI® *PMP Examination Content Outline,* 2015, Initiating, 5, Task 1

2. a. Reviewing, approving, and controlling changes

   Performing integrated change control consists of coordinating and managing changes across the project. Activities that occur within the context of Performance Integrated Change Control include: validate scope, control scope, control schedule, control costs, perform quality assurance, control quality, manage the project team, control communications, control risks, conduct procurements, control procurements, manage stakeholder engagement, and control stakeholder engagement. [Monitoring and Controlling]

   PMI®, *PMBOK® Guide*, 2013, 94–95
   PMI® *PMP Examination Content Outline,* 2015, Monitoring and Controlling, 9, Task 1

3. b. Not mix the types of meetings on your project

   Meetings are a tool and technique used in Direct and Manage Project Work. Meetings tend to be one of three types: information exchange; brainstorming, option evaluation, or design; or decision making. A best practice is to not combine the types of meetings and prepare for them with a well-defined agenda, purpose, objective, and time frame. They should be documented using minutes and action items. [Executing]

   PMI®, *PMBOK® Guide*, 2013, 84
   PMI® *PMP Examination Content Outline,* 2015, Executing, 8, Task 6

4. a. Project charter

   The project charter documents the business needs, assumptions, constraints, understanding of the customer needs and high-level requirements and what the new product, service, or result is to satisfy. It is the document used to formally authorize the project. [Initiating]

   PMI®, *PMBOK® Guide*, 2013, 71–72
   PMI® *PMP Examination Content Outline,* 2015, Initiating, 5, Task 5

5. d. Application area and complexity

   The content of the project management plan is primarily influenced by the application area [in this case drug development] and complexity of the project. The size of the plan is typically commensurate with the size and complexity of the project. [Planning]

   PMI®, *PMBOK® Guide*, 2013, 74
   PMI® *PMP Examination Content Outline,* 2015, Planning, 6, Task 1

6. d. Project management plan

   The project management plan must be updated with changes to subsidiary plans and baselines subject to formal change control processes. Those changes must be communicated to appropriate stakeholders in a timely manner. [Monitoring and Controlling]

   PMI®, *PMBOK® Guide*, 2013, 100
   PMI® *PMP Examination Content Outline,* 2015, Monitoring and Controlling, 9, Task 2

7. b. Project charter

   Although the project charter cannot stop conflicts from arising, it can provide a framework to help resolve them, because it describes the project manager's authority to apply organizational resources to project activities. In addition it documents the business needs, assumptions, constraints, an understanding of the customer's needs and high-level requirements, and the project's new product, service, or result. [Initiating]

   Meredith, J.R. and Mantel, Jr., S.J., *Project Management: A Managerial Approach*, 2012, 228
   PMI®, *PMBOK® Guide*, 2013, 71–72
   PMI®*PMP Examination Content Outline,* 2015, Initiating, 5, Task 5

8. d. Evaluate the return on capital percent versus the cost of capital percent

   EVA quantifies the value a company provides to its investors and seeks to determine if a company is creating or destroying value to its shareholders. It is calculated by subtracting the expected return, (represented by the capital charge), from the actual return that a company generates, (represented by net operating profit after taxes). As a decision is made to determine whether or not to initiate a project based on its business case, a variety of cost-benefit analyses typically are undertaken, and EVA is one that is extremely useful. [Initiating]

   Stewart B.G, *Best Practice-EVA: The Definitive Guide to Measuring and Maximizing Stakeholder Value,* 2013, p. 300
   PMI®, *PMBOK® Guide*, 2013, 69
   PMI® *PMP Examination Content Outline,* 2015, Initiating, 5, Task 7

9. c. Meeting management

   Meeting management is an example of a facilitation technique used in developing the project charter as meetings may be held with key stakeholders and subject matter experts. Other facilitation techniques used to guide preparation of the charter are brainstorming, problem solving, and conflict resolution. [Initiating]

   PMI®, *PMBOK® Guide*, 2013, 71
   PMI® *PMP Examination Content Outline,* 2015, Initiating, 5, Task 5

10. b. Provide overall management of the project work

    While all of the answers apply to the direct and manage project work process, the key benefit is that it involves providing overall management of the work of the project, encompassing the other answers listed. In this question, the key words were “the key benefit”. Had the question instead asked about the purpose of the process, answer c would have been the correct choice. [Executing]

    PMI®, *PMBOK® Guide*, 2013, 79
    PMI® *PMP Examination Content Outline,* 2015, Executing, 8, Task 2

11. a. Guidelines and criteria

    While you are managing a different type of project, the organization has managed projects before and therefore may have as part of its organizational process assets a project management plan template, which sets forth guidelines and criteria to tailor the organization's processes to satisfy specific needs of the project. In addition, project closure guidance or requirements, such as the product validation and acceptance criteria, may be part of organizational process assets for use in the template for the project management plan. [Planning]

    PMI®, *PMBOK® Guide*, 2013, 75
    PMI® *PMP Examination Content Outline,* 2015, Planning, 6, Task 11

12. d. Project statement of work

    The project statement of work is a useful document as it describes the products, services, or results the project is to deliver. It references the business need, product scope description, and the strategic plan. While the question relates to planning, the statement of work also is used to develop the project's charter. In planning, it is listed one of many project documents that are not part of the project management plan. [Initiating and Planning]

    PMI®, *PMBOK® Guide*, 2013, 68, 78
    PMI® *PMP Examination Content Outline,* 2015, Initiating, 5, Task 2
    PMI® *PMP Examination Content Outline,* 2015, Planning, 6, Task 1

13. a. Standardized guidelines and work instructions

    Standardized guidelines and work instructions are an organizational process asset to consider as the project management plan is developed. They include guidelines and criteria to tailor the organization's set of standard processes to satisfy the specific needs of the project. [Planning]

    PMI®, *PMBOK® Guide*, 2013, 75
    PMI® *PMP Examination Content Outline,* 2015, Planning, 6, Task 11

14. b. Used to review, evaluate, approve, delay, or reject changes to the project

    The change control board's powers and responsibilities should be well defined and agreed upon by key stakeholders. On some projects, multiple change control boards may exist with different areas of responsibility. It is a formally chartered group, and in addition to the items in the answer, it also records and communicates decisions that are made. The level of change control depends on the application area and project complexity as well as the project's environment. On some change control boards, customer and/or sponsor approval may be required for certain change requests, unless a customer representative or a sponsor is a board member. [Monitoring and Controlling]

    PMI®, *PMBOK® Guide*, 2013, 96
    PMI® *PMP Examination Content Outline,* 2015, Monitoring and Controlling, 9, Task 2

15. b. Project management information systems

    The items listed are part of these systems, a tool and technique in both executing and monitoring and control processes. In executing, it is considered part of environmental factors and also includes a work authorization system, a configuration management system, and an information collection and distribution system along with interfaces to other automated systems. Automated gathering and reporting on key performance indicators can be part of the PMIS. In monitoring and controlling, it also includes scheduling, cost, and resourcing tools in addition to the items in the answer. [Executing and Monitoring and Controlling]

    PMI®, *PMBOK® Guide*, 2013, 84, 92
    PMI® *PMP Examination Content Outline,* 2015, Executing, 8, Task 2
    PMI® *PMP Examination Content Outline,* 2015, Monitoring and Controlling, 9, Task 1

16. b. Updates

    Change requests may include corrective actions, preventive actions, defect repairs, or updates. Updates are changes to formally controlled project documents or plans to reflect modified or additional content. [Executing]

    PMI®, *PMBOK® Guide*, 2013, 85
    PMI® *PMP Examination Content Outline,* 2015, Executing, 8, Task 4

17. a. Procedures that define how project documents may be changed

A change control system is a collection of formal, documented procedures that define the process used to control change and approve or reject changes to project documents, deliverables, or baselines. It includes the paperwork, tracking systems, and approval levels necessary to authorize changes. Changes may be requested by any stakeholder involved with the project. The goal is to review all change requests, approve and manage changes to deliverables and organizational process assets, project documents, and the project management plan, and communicate their disposition. [Monitoring and Controlling]

PMI®, *PMBOK® Guide*, 2013, 94, 96
PMI® *PMP Examination Content Outline,* 2015, Monitoring and Controlling, 9, Task 1

18. d. Project management plan

The project management plan describes how the project will be executed and monitored and controlled. While it contains a number of subsidiary plans, it also contains other items including information on key management reviews for contents, their extent, and timing to address open issues and pending decisions. [Planning]

PMI®, *PMBOK® Guide*, 2013, 76–77
PMI® *PMP Examination Content Outline,* 2015, Planning, 6, Task 11

19. d. Executing

Executing is where the majority of the budget is spent because this is the process where all of the resources (people, material, etc.) are applied to the activities and tasks in the project management plan. It also involves coordinating people and other resources, managing stakeholder expectations, and integrating and performing the work described in the project management plan. [Executing]

PMI®, *PMBOK® Guide*, 2013, 56
PMI® *PMP Examination Content Outline,* 2015, Executing, 8, Task 2

20. a. Project statement of work

The project statement of work describes in a narrative form the products, services, or results that the project will deliver. For external projects, it can be received from the customer as part of the bid document. It references the product scope description, business need, and the strategic plan. [Initiating]

PMI®, *PMBOK® Guide*, 2013, 68
PMI® *PMP Examination Content Outline,* 2015, Initiating, 5, Task 2

21. c. For successful completion of the current and subsequent project phases

    Rolling wave planning provides progressive detailing of the work to be accomplished throughout the life of the project, indicating that planning and documentation are iterative and ongoing processes. Through this approach, where a more general and high-level plan is available, more detailed planning is executed for appropriate time windows as new work activities are to begin and resources are to be assigned. The work in the near term then is planned. In detail, while future work is planned at a higher level. It is especially useful on longer projects. [Planning]

    PMI®, *PMBOK® Guide*, 2013, 45 and 560
    PMI® *PMP Examination Content Outline,* 2015, Planning, 6, Task 11

22. d. Ensuring that project scope changes are reflected in changes to product scope

    Integrated change control requires maintaining the integrity of baselines by releasing only approved changes into project products, services, or results. It also ensures that changes to product scope are reflected in the project scope definition. This is done by coordinating changes across the entire project. The project management plan, project scope statement, and other deliverables are maintained by carefully and continuously managing changes and in doing so by ensuring only approved changes are incorporated into a revised baseline. [Monitoring and Controlling]

    PMI®, *PMBOK® Guide*, 2013, 94, 96, and 99–100
    PMI® *PMP Examination Content Outline,* 2015, Monitoring and Controlling, 9, Task 2

23. b. Does not apply to the product to be created

    A project is completed when its objectives have been achieved or when they are recognized as being unachievable, and the project is terminated, or when the need for the project no longer exists. Thus, the concept of *temporary* applies to the project life cycle—not the product life cycle. Further, temporary does not necessarily mean the project's duration is short as it refers to the project's engagement and its longevity. It also does not typically apply to the product, service, or result as most projects are undertaken to create a lasting outcome with social, economic, or environmental impacts that far outlive the project. [Planning]

    PMI®, *PMBOK® Guide*, 2013, 3–4
    PMI® *PMP Examination Content Outline,* 2015, Planning, 6, Task 1

24. d. Review the scope baseline

In closing the project, it is necessary to ensure that the project work is completed, and the project has met its objectives. Since project scope is measured against the project management plan, the project manager then reviews the scope baseline to ensure completion. [Closing]

PMI®, *PMBOK® Guide*, 2013, 101
PMI® *PMP Examination Content Outline,* 2015, Closing, 11, Task 3

25. a. Technical

Scope, time, and cost are examples of project baselines to be part of the project management plan. A technical baseline would be covered by the scope baseline.

PMI®, *PMBOK® Guide*, 2013, 76
PMI® *PMP Examination Content Outline,* 2015, Planning, 6, Task 11

26. a. Employees who can apply the training effectively

Most deliverables are tangible, such as buildings or roads, but intangible deliverables also can be provided and may be even more important. Work performance data are collected during direct and manage project work and is passed on to the monitoring and controlling processes of each process area for further analysis. [Executing]

PMI®, *PMBOK® Guide*, 2013, 84–85
PMI® *PMP Examination Content Outline,* 2015, Executing, 8, Task 2

27. b. Corrective action

When you volunteered, you signed a confidentiality statement so you could not disclose what was under way on this activity. Now the Standard has been issued, and to stay in alignment with PMI®'s best practices, you need to issue a change request based on corrective action to realign the performance of the work of your project with your project management plan. [Executing]

PMI®, *PMBOK® Guide*, 2013, 85
PMI® *PMP Examination Content Outline,* 2015, Executing, 8, Task 4

28. a. During the initiating processes

When the project manager is selected and assigned to the project during initiation, several of the usual start-up tasks for a project are simplified. In addition, becoming involved with project activities from the beginning helps the project manager to understand where the project fits within the organization in terms of its priority relative to other projects and the ongoing work of the organization. Further, if assigned early, the project manager can participate in development of the project charter. [Initiating]

Meredith, J.R. and Mantel, Jr., S.J., *Project Management: A Managerial Approach*, 2012, 101
PMI®, *PMBOK® Guide*, 2013, 67
PMI® *PMP Examination Content Outline,* 2015, Initiating, 5, Task 6

29. a. Useful information may be lost

Closure includes collecting project records, ensuring that the records accurately reflect final specifications, analyzing project or phase success and effectiveness, and archiving such information for future use. Each phase of the project should be properly closed while important project information is still available. A key benefit is it provides lessons learned, the formal ending of project work, and the release of resources to pursue new endeavors. [Closing]

PMI®, *PMBOK® Guide*, 2013, 100–101
PMI® *PMP Examination Content Outline,* 2015, Closing, 11, Task 5

30. c. Work performance data

Work performance data are the raw observations and measurements identified during activities performed to carry out the work of the project. Other examples are the reported percent of work physically completed, quality and technical performance measures, number of defects, actual costs, and actual durations. It is important to note that the PMBOK® recognizes the terms data and information are used interchangeably in practice, which can lead to confusion among stakeholders. It has therefore set up a classification system of work performance data, work performance information, and work performance reports. [Executing]

PMI®, *PMBOK® Guide*, 2013, 59
PMI® *PMP Examination Content Outline,* 2015, Executing, 8, Task 6

31. a. Overlap and interact throughout the project

Project management processes and product-oriented management processes must be integrated throughout the project's life cycle, given their close relationship. In some cases, it is difficult to distinguish between the two. For example, knowing how the product will be created aids in determining the project's scope. However, the project life cycle is independent from the product that is produced or modified by the project. The project should take the current life-cycle phase of the product into consideration. [Executing]

PMI®, *PMBOK® Guide,* 2013, 38–39
PMI® *PMP Examination Content Outline,* 2015, Executing, 8, Task 2

32. d. Documentation that completed deliverables have been accepted

Documentation that the completed deliverables have been accepted is prepared as an output of validate scope, which then is forwarded to the close project or phase process. The close project or phase procedures provide a listing of necessary activities, including: confirmation that the project has met sponsor, customer, and other stakeholder requirements; satisfaction and validation that the completion and exit criteria have been met; the transfer of deliverables to the next phase or to production/operations has been accomplished; and activities to collect, audit, and archive project information and gather lessons learned have been addressed. [Closing]

PMI®, *PMBOK® Guide,* 2013, 100–103
PMI® *PMP Examination Content Outline,* 2015, Closing, 11, Task 1

33. c. A sponsor

The project charter should be issued by a project initiator or sponsor who formally authorizes the project and provides the project manager with the authority to apply organizational resources to project activities. The sponsor may be a person or a group and is accountable for enabling success, providing resources and support for the project. The project charter should not be issued by the project manager, although, the project manager can assist in its development. [Initiating]

PMI®, *PMBOK® Guide,* 2013, 71
PMI® *PMP Examination Content Outline,* 2015, Initiating, 5, Task 6

34. c. Prepare a change request

The change request should detail the nature of the change and its effect on the project. Documentation is critical to provide a record of the change and who approved it, in case differences of opinion arise later. A change request is an output from the direct and manage project work process and an input to the perform integrated change control process It serves to modify any document, deliverable, or baseline, Change requests can be direct or indirect, externally or internally initiated, and can be optional or legally or contractually mandated. It should be noted that while change requests may include corrective action, preventive action, defect repair, and updates, corrective and preventive actions do not normally affect the project baselines, only the performance against the baselines. [Executing and Monitoring and Controlling]

PMI®, *PMBOK® Guide*, 2013, 85, 97
PMI® *PMP Examination Content Outline,* 2015, Executing, 8, Task 4
PMI® *PMP Examination Content Outline,* 2015, Monitoring and Controlling, 9, Task 2

35. d. More rework

The question is an example of an overlapping relationship between phases. It is used to compress the schedule through fast tracking as an example. By overlapping phases more resources may be needed, risks may increase, and more rework may result if a significant phase progresses before accurate information is available from the previous phase. It should be noted that there is no ideal structure that applies to all projects; some only have one phase, while other projects even in the same organization may have several phases. Overlapping relationships and sequential relationships are the two basic types of phase-to-phase relationships. In contrast, in the sequential approach, a phase starts only after the previous phase is complete. This sequential approach reduces uncertainty but may eliminate options for reducing the overall schedule. [Executing]

PMI®, *PMBOK® Guide*, 2013, 42–43
PMI® *PMP Examination Content Outline,* 2015, Executing, 8, Task 2

36. b. Predictive

    A predictive life cycle or one that is fully plan driven is recommended if the product to be delivered is well understood, there is a substantial base of industry practice, or if a product is required to be delivered in full to have value to stakeholder groups. The project's scope, time, and cost to deliver it are determined in the project life cycle as early as possible. In this type of life cycle, projects proceed through a series of sequential or overlapping phases, with each phase focusing on a subset of project activities and project management processes. [Planning]

    PMI®, *PMBOK® Guide*, 2013, 44–45
    PMI® *PMP Examination Content Outline,* 2015, Planning, 6, Task 12

37. c. Conduct an immediate review of the work packages

    A thorough review of the work packages will provide a complete accounting of the physical progress achieved on the project. This is the first step in attempting to improve performance. The project manager reviews prior information, investigates and documents the reasons for actions taken if the project is terminated before completion, and engages stakeholders as appropriate in the process. [Closing]

    Cleland, D. and Ireland, L., *Project Manager's Handbook*, 2007, 365–375
    PMI®, *PMBOK® Guide*, 2013, 101
    PMI® *PMP Examination Content Outline,* 2015, Closing, 11, Task 5

38. c. There are misunderstandings regarding what was agreed upon by stakeholders

    Using a formal, documented approach to change management reduces the level of misunderstanding or uncertainty regarding the nature of the change and its impact on cost and schedule. For large projects, change control boards are recommended. Another key benefit is it allows for documented changes within the project to be considered in an integrated fashion, which helps to reduce risks if changes are made without consideration to the project's overall objectives or plans. [Monitoring and Controlling]

    Meredith, J.R. and Mantel, Jr., S.J., *Project Management: A Managerial Approach*, 2012, 500
    PMI®, *PMBOK® Guide*, 2013, 94
    PMI® *PMP Examination Content Outline,* 2015, Monitoring and Controlling, 9, Task 2

39. b. There are new stakeholders.

As long as the new stakeholders agree with the project's business case and have little or limited influence with the project, the work should continue. However, if any of the other events occur, termination should be considered. Ideally, the organization has project closure guidelines that include the necessary steps to take should termination be necessary. Also recommended is formal documentation that states why termination was needed and is part of the project closure documents. [Closing]

Cleland, D. and Ireland, L., *Project Manager's Handbook,* 2007, 365–375
PMI®, *PMBOK® Guide,* 2013, 101, 104
PMI® *PMP Examination Content Outline,* 2015, Closing, 11, Task 3

40. b. Configuration management plan

A configuration management plan is part of a project management plan to document how configuration management will be performed on the project. It documents configuration identification, configuration status accounting, and configuration verification and audit activities that will be performed as well as whether change control boards will be used. [Planning]

PMI®, *PMBOK® Guide,* 2013, 77, 96–97
PMI® *PMP Examination Content Outline,* 2015, Planning, 6, Task 11

Project Scope Management

Study Hints

The Project Scope Management questions on the PMP® certification exam cover a diverse, yet fundamental, set of project management topics. Planning scope management activities, collecting requirements, defining requirements with a scope statement, creating the work breakdown structure (WBS), validating the scope with accepted deliverables, and managing scope changes are among the topics covered.

PMI® views scope management as a six-step processes that consists of: plan scope management, collect requirements, define scope, create WBS, validate scope, and control scope. *PMBOK® Guide* Figure 5-1 provides an overview of the structure of Project Scope Management. Know this chart thoroughly.

The Project Scope Management questions on the exam are straightforward. Historically, most people have found these questions to be relatively easy; however, do not be lulled into a false sense of security by past results. These questions cover a wide breadth of material, and you must be familiar with the terminology and perspectives adopted by PMI®.

You also may wish to consult PMI®'s *Practice Standard for Work Breakdown Structures*—Second Edition (2006) for additional information. Its *Requirements Guide* (2013) is another great resource along with its *Practice Standard on Configuration Management* (2007).

Following is a list of the major Project Scope Management topics. Use it to help focus your study efforts on the areas most likely to appear on the exam.

Major Topics

Plan Scope Management

- Preparing a scope management plan
- Preparing a requirements management plan

Collect requirements

- Tools and techniques
- Requirements documentation
- Requirements traceability matrix

Define scope

- Product analysis
- Alternatives analysis
- Facilitated workshops
- Project scope statement

Create WBS

- Benefits
- Uses
- Development/decomposition
- WBS dictionary
- Scope baseline

Validate scope

- Inspection
- Group decision-making techniques
- Accepted deliverables
- Change requests
- Work performance information

Control scope

- Variance analysis
- Work performance information
- Updates to the project management plan and to the project documents

Practice Questions

INSTRUCTIONS: Note the most suitable answer for each multiple-choice question in the appropriate space on the answer sheet.

1. Progressive elaboration of product characteristics on your project must be coordinated carefully with the—

   a. Proper project scope definition
   b. Project stakeholders
   c. Scope change control system
   d. Customer's strategic plan

2. You are examining multiple scope change requests on a project you were asked to take over because the previous project manager decided to resign. To assess the degree to which the project scope will change, you need to compare the requests to which project document?

   a. Preliminary scope statement
   b. WBS
   c. Change management plan
   d. Scope management plan

3. You and your project team recognize the importance of project scope management to a project's overall success; therefore, you include only the work required for successful completion of the project. The first step in the Project Scope Management process is to—

   a. Clearly distinguish between project scope and product scope
   b. Prepare a scope management plan
   c. Define and document your stakeholders' needs to meet the project's objectives
   d. Capture and manage both project and product requirements

4. An example of an organizational process asset that could affect how project scope is to be managed is—

   a. Personnel administration
   b. Marketplace conditions
   c. Historical information
   d. Organizational culture

5. You are managing a complex project for a new method of heating and air conditioning in vehicles. You will use both solar and wind technologies in this project to reduce energy costs. Therefore, you must ensure that the work of your project will result in delivering the project's specified scope, which means that you should measure completion of the product scope against the—

    a. Scope management plan
    b. Project management plan
    c. Product requirements
    d. Requirements management plan

6. A key tool and technique used in Define Scope is—

    a. Templates, forms, and standards
    b. Decomposition
    c. Expert judgment
    d. Project management methodology

7. Alternatives generation often is useful in defining project scope. An example of a technique that can be used is—

    a. Sensitivity analysis
    b. Decision trees
    c. Mathematical model
    d. Lateral thinking

8. Product analysis techniques include all the following EXCEPT—

    a. Value engineering
    b. Value analysis
    c. Systems analysis
    d. Bill of materials

9. The baseline for evaluating whether requests for changes or additional work are contained within or outside the project's exclusion is provided by the—

    a. Project management plan
    b. Project scope statement
    c. Project scope management plan
    d. WBS dictionary

10. Rather than use a WBS, your team developed a bill of materials to define the project's work components. A customer review of this document uncovered that a scope change was needed, because a deliverable had not been defined, and a change request was written subsequently. This is an example of a change request that was the result of—

    a. An external event
    b. An error or omission in defining the scope of the product
    c. A value-adding change
    d. An error or omission in defining the scope of the project

11. Collecting requirements is critical in project scope management as it becomes the foundation for the project's—

    a. Scope management plan
    b. WBS
    c. Schedule
    d. Scope change control system

12. The project scope statement addresses and documents all the following items EXCEPT—

    a. Project exclusions
    b. The relationship between the deliverables and the business need
    c. Product scope description
    d. Project management methodology (PMM)

13. The first step in collecting requirements on any project, large or small, is to—

    a. Talk with the project stakeholders through interviews
    b. Review the scope management plan
    c. Conduct facilitated workshops with stakeholders
    d. Prepare a requirements document template that you and your team can use throughout the collect requirements process

14. You want to structure your project so that each project team member has a discrete work package to perform. The work package is a—

    a. Deliverable at the lowest level of the WBS
    b. Task with a unique identifier
    c. Required level of reporting
    d. Task that can be assigned to more than one organizational unit

15. Quality function deployment is one approach for collecting requirements. Assume that you have studied the work of numerous quality experts, such as Deming, Juran, and Crosby, and your organization has a policy that states the importance of quality as the key constraint of all project constraints. You and your team have decided to use quality function deployment on your new project to manufacture turbines that use alternative fuels. The first step you should use is to—

    a. Determine the voice of the customer
    b. Build the house of quality
    c. Address the functional requirements and how best to meet them
    d. Hold a focus group of prequalified stakeholders

16. On the WBS, the first level of decomposition may be displayed by using all the following EXCEPT—

    a. Phases of the project life cycle
    b. Subcomponents
    c. Major deliverables
    d. Project organizational units

17. Change is inevitable on projects. Uncontrolled changes are often referred to as—

    a. Rework
    b. Scope creep
    c. Configuration items
    d. Emergency changes

18. Each WBS component should be assigned a unique identifier from a code of accounts to—

    a. Link the WBS to the bill of materials
    b. Enable the WBS to follow a similar numbering system to that of the organization's units as part of the organizational breakdown structure
    c. Sum costs, schedule, and resource information
    d. Link the WBS to the project management plan

19. In scope control it is important to determine the cause of any unacceptable variance relative to the scope baseline. This can be done through—

    a. Root-cause analysis
    b. Control charts
    c. Inspections
    d. Project performance measurements

20. To assist your software development team in collecting requirements from potential users and to ensure that agreement about the stakeholders' needs exists early in the project, you decide to use a group creativity technique. Numerous techniques are available, but you and your team choose a voting process to rank the most useful ideas for further prioritization. This approach is known as—

    a. Brainstorming
    b. Nominal group technique
    c. Delphi technique
    d. Affinity diagram

21. You have been appointed project manager for a new project in your organization and must prepare a project management plan. You decide to prepare a WBS to show the magnitude and complexity of the work involved. No WBS templates are available to help you. To prepare the WBS, your first step should be to—

    a. Determine the cost and duration estimates for each project deliverable
    b. Identify and analyze the deliverables and related work
    c. Identify the components of each project deliverable
    d. Determine the key tasks to be performed

22. You want to avoid scope creep on your project and are working hard to do so. Your sponsor has asked for regular reports as to how the project is performing according to the scope baseline. You should provide the sponsor with which one of the following—

    a. Variance analysis
    b. Inspection and peer review results
    c. Work performance information
    d. The impact analysis results of proposed change requests

23. You are leading a project team to identify potential new products for your organization. One idea was rejected by management because it would not fit with the organization's core competencies. You need to recommend other products using management's guideline as—

    a. An assumption
    b. A risk
    c. A specification
    d. A technical requirement

24. Validate scope—

    a. Improves cost and schedule accuracy, particularly on projects using innovative techniques or technology
    b. Is the last activity performed on a project before handoff to the customer
    c. Documents the characteristics of the product, service, or result that the project was undertaken to create
    d. Differs from perform quality control in that validate scope is concerned with the acceptance—not the correctness—of the deliverables

25. Any step recommended to bring expected future performance in line with the project management plan is called—

    a. Performance evaluation
    b. Corrective action
    c. Preventive action
    d. Defect repair

26. One approach that can be used to detect the impact of any change from the scope baseline on the project objectives is—

    a. Requirements traceability matrix
    b. Review of formal and informal scope control processes and guidelines
    c. A formal configuration management plan
    d. Well-documented requirements

27. Updates of organizational process assets that are an output of Control Scope include all the following EXCEPT—

    a. Causes of variations
    b. Lessons learned
    c. Work authorization system
    d. Reasons certain corrective actions were chosen

28. Work performance information in Validate Scope includes all the following EXCEPT—

    a. Started deliverables
    b. Costs authorized and incurred
    c. Progress of deliverables
    d. Completed deliverables

29. Your project is now under way, and you are working with your team to prepare your requirements management plan. Which of the following strongly influences how requirements are managed?

    a. The phase-to-phase relationship
    b. A set of procedures by which project scope and product scope may be changed
    c. Requirements traceability matrix
    d. Requirements documentation

30. You are the project manager on a systems engineering project designed to last six years and to develop the next-generation corvette for use in military operations. You and your team recognize that requirements may change as new technologies, especially in sonar systems, are developed. You are concerned that these new technologies may lead to changes in the scope of your product, which then will affect the scope of your project. Therefore your requirements traceability matrix should include tracing requirements to all the following project elements EXCEPT—

    a. Business needs
    b. Product design
    c. Product development
    d. Project verification

31. Your customer signed off on the requirements document and scope statement of your video game project last month. Today she stated she would like to make it an interactive game that can be played on a television and on a computer. This represents a requested scope change that, at a minimum—

    a. Should be reviewed according to the Perform Integrated Change Control process
    b. Results in a change to all project baselines
    c. Requires adjustments to cost, time, quality, and other objectives
    d. Results in a lesson learned

32. The key inputs to the Validate Scope process include all the below items EXCEPT—

    a. The project management plan (scope management plan and scope baseline)
    b. Change requests
    c. Verified deliverables
    d. Requirements traceability matrix

33. Modifications may be needed to the WBS and WBS dictionary because of approved change requests, which shows that—
    a. Re-planning is an output of Control Scope
    b. Scope creep is common on projects
    c. Re-baselining will be necessary
    d. Updates are needed to the scope baseline

34. You and your team are documenting requirements on you project to control fatigue as people need to work more hours to keep up with the competition. You decided to set up components for the requirements on your project. Acceptance criteria are an example of—
    a. Stakeholder requirements
    b. Transition requirements
    c. Project requirements
    d. Business requirements

35. Which following item is NOT an input to Control Scope?
    a. Requirements traceability matrix
    b. Work performance information
    c. Verified deliverables
    d. Scope management plan

36. You are the project manager for a subcontractor on a major contract. The prime contractor has asked that you manage your work in a detailed manner. Your first step is to—
    a. Follow the WBS that the prime contractor developed for the project and use the work packages you identified during the proposal
    b. Develop a subproject WBS for the work package that is your company's responsibility
    c. Establish a similar coding structure to the prime contractor's to facilitate use of a common project management information system
    d. Develop a WBS dictionary to show specific staff assignments

37. The project scope statement is important in scope control because it—
    a. Is a critical component of the scope baseline
    b. Provides information on project performance
    c. Alerts the project team to issues that may cause problems in the future
    d. Is expected to change throughout the project

38. The product scope description is documented as part of the project's scope statement. It is important to include it because it—

    a. Facilitates the project acceptance process
    b. Describes specific constraints associated with the project
    c. Progressively elaborates characteristics
    d. Shows various alternatives considered

39. How is a context diagram used?

    a. To depict product scope
    b. To trace requirements as part of the traceability matrix
    c. To develop the scope management plan
    d. To develop the requirements management plan

40. You are establishing a PMO that will have a project management information system that will be an online repository of all program data. You will collect descriptions of all work components for each project under the PMO's jurisdiction. This information will form an integral part of the—

    a. Chart of accounts
    b. WBS dictionary
    c. WBS structure template
    d. Earned value management reports

Answer Sheet

1.	a	b	c	d	21.	a	b	c	d
2.	a	b	c	d	22.	a	b	c	d
3.	a	b	c	d	23.	a	b	c	d
4.	a	b	c	d	24.	a	b	c	d
5.	a	b	c	d	25.	a	b	c	d
6.	a	b	c	d	26.	a	b	c	d
7.	a	b	c	d	27.	a	b	c	d
8.	a	b	c	d	28.	a	b	c	d
9.	a	b	c	d	29.	a	b	c	d
10.	a	b	c	d	30.	a	b	c	d
11.	a	b	c	d	31.	a	b	c	d
12.	a	b	c	d	32.	a	b	c	d
13.	a	b	c	d	33.	a	b	c	d
14.	a	b	c	d	34.	a	b	c	d
15.	a	b	c	d	35.	a	b	c	d
16.	a	b	c	d	36.	a	b	c	d
17.	a	b	c	d	37.	a	b	c	d
18.	a	b	c	d	38.	a	b	c	d
19.	a	b	c	d	39.	a	b	c	d
20.	a	b	c	d	40.	a	b	c	d

Answer Key

1. a. Proper project scope definition

   Progressive elaboration of a project's specification must be coordinated carefully with proper scope definition, particularly when the project is performed under contract. When properly defined, the project scope—the work to be done—should remain constant even when the product characteristics are elaborated progressively. [Planning]

   PMI®, *PMBOK® Guide*, 2013, 6, 105–106
   PMI® *PMP Examination Content Outline,* 2015, Planning, 6, Task 2

2. b. WBS

   The WBS, along with the detailed scope statement and the WBS dictionary, defines the project's scope baseline, which provides the basis for any changes that may occur on the project. Formal change control procedures should be used if the scope baseline requires change. On the exam, recognize how the WBS is developed and the definitions of a work package, a control account, and a planning package in case one appears in the exam. [Planning]

   PMI®, *PMBOK® Guide*, 2013, 131–132
   PMI® *PMP Examination Content Outline,* 2015, Planning, 6, Task 2

3. b. Prepare a scope management plan

   The work involved in the six Project Scope Management processes begins by preparing a scope management plan, which is a subsidiary plan for the project management plan. It documents how project scope is to be defined, validated, and controlled. It also describes the Project Scope Management processes from definition to control and provides guidance as to how scope will be managed throughout the project. It contains processes to prepare a project scope statement and then to prepare the WBS, how the WBS will be maintained and approved, how formal acceptance of the deliverables will be done, and how requests for changes will be controlled. [Planning]

   PMI®, *PMBOK® Guide*, 2013, 106–107, 109–110
   PMI® *PMP Examination Content Outline,* 2015, Planning, 6, Task 2

4. c. Historical information

   In addition to the answer, organizational process assets that can influence plan scope management include formal and informal policies, procedures, and guidelines impacting project scope management. The lessons-learned knowledge base is another example of an organizational process asset. [Planning]

   PMI®, *PMBOK® Guide*, 2013, 109
   PMI® *PMP Examination Content Outline,* 2015, Planning, 6, Task 2

5. c. Product requirements

   Completion of the project scope is measured against the project management plan, and completion of the product scope is measured against the requirements. In the project context, product scope consists of features and functions that characterize the product, service, or result. Project scope is the work that must be done to deliver the product, service, or result with specified features and functions. [Planning]

   PMI®, *PMBOK® Guide*, 2013, 105–106
   PMI® *PMP Examination Content Outline,* 2015, Planning, 6, Task 1

6. c. Expert judgment

   Expert judgment is used to analyze the information needed to develop a project scope statement. It is applied to any technical details. Expert judgment can be provided by any group or individual with specialized knowledge or training such as other parts of the organization, consultants, stakeholders, professional and technical associations, industry groups, and subject matter experts. [Planning]

   PMI®, *PMBOK® Guide*, 2013, 122
   PMI® *PMP Examination Content Outline,* 2015, Planning, 6, Task 2

7. d. Lateral thinking

   Lateral thinking, brainstorming, and analysis of alternatives are examples of alternatives generation that can be used to develop as many potential options as possible to execute and perform the project's work. Lateral thinking first was used in 1967 by Edward de Bono and refers to the ability to think 'outside the box', using inspiration, and creativity. [Planning]

   PMI®, *PMBOK® Guide*, 2013, 123
   PMI® *PMP Examination Content Outline,* 2015, Planning, 6, Task 2

8. d. Bill of materials

   Product analysis techniques vary by application area, and each application area generally has accepted methods to translate project objectives into tangible deliverables and requirements. These techniques are used for those products with a product as a deliverable. Other product analysis techniques include product breakdown, requirements analysis, requirements analysis, systems engineering, value analysis, and value engineering. [Planning]

   PMI®, *PMBOK® Guide*, 2013, 122
   PMI® *PMP Examination Content Outline,* 2015, Planning, 6, Task 2

9. b. Project scope statement

   Project exclusion identifies generally what is excluded from the project. These scope exclusions help manage stakeholder expectations. It enables the project manager and team to perform more detailed planning, guides work during execution, and provides the baseline to evaluate whether requests for changes or additional work are contained within or outside of the project's boundaries. [Planning]

   PMI®, *PMBOK® Guide*, 2013, 123
   PMI® *PMP Examination Content Outline,* 2015, Planning, 6, Task 13

10. b. An error or omission in defining the scope of the product

    The bill of materials provides a hierarchical view of the physical assemblies, subassemblies, and components needed to build a manufactured product, whereas the WBS is a deliverable-oriented grouping of project components used to define the total scope of the project, providing a structured vision of what has to be delivered. Using a bill of materials where a WBS would be more appropriate may result in an ill-defined scope and subsequent change requests. These change requests may include preventive or corrective action, defect repairs, or enhancement requests. [Monitoring and Controlling]

    PMI®, *PMBOK® Guide*, 2013, 125–126, 140
    PMI® *PMP Examination Content Outline,* 2015, Monitoring and Controlling, 9, Task 2

11. b. WBS

Collecting requirements provides the basis for defining project scope and product scope. It also involves determining, documenting, and managing stakeholder needs to meet project objectives. Requirements documentation is an input to the Create WBS process as detailed requirement documentation is essential to understand what needs to be produced by the project and what needs to be done to deliver the project and its final products. The requirements become the foundation for the WBS; moreover, cost, schedule, and quality planning are built upon the requirements, and the requirements management plan is a component of the project management plan. [Planning]

PMI®, *PMBOK® Guide*, 2013, 110, 127
PMI® *PMP Examination Content Outline,* 2015, Planning, 6, Task 2

12. d. Project management methodology (PMM)

The PMM is an organization-approved approach for project management that is used on every project in some organizations. It is not part of the project scope statement, which describes the project scope description deliverables, acceptance criteria, project exclusions, assumptions, and constraints. The project scope statement documents the entire scope, including product and project scope and describes the project's deliverables and the work required to complete them. It provides a common understanding of the project's scope among stakeholders. [Planning]

PMI®, *PMBOK® Guide*, 2013, 123–124
PMI® *PMP Examination Content Outline,* 2015, Planning, 6, Task 2

13. b. Review the scope management plan

The scope management plan is reviewed first as it provides clarity as to how the project team will determine which requirements need to be collected on the project. The next step is to work to document the requirements and prepare a requirements traceability matrix. [Planning]

PMI®, *PMBOK® Guide*, 2013, 113, 117–119
PMI® *PMP Examination Content Outline,* 2015, Planning, 6, Task 13

14. a. Deliverable at the lowest level of the WBS

    A work package is the lowest or smallest unit of work division in a project or WBS. The work package can be used to group activities where work is scheduled and estimated, monitored, and controlled. Recognize that in the context of the WBS, work refers to work products or deliverables that are the result of activity, not the activity itself. [Planning]

    PMI®, *PMBOK® Guide*, 2013, 126
    PMI® *PMP Examination Content Outline,* 2015, Planning, 6, Task 2

15. a. Determine the voice of the customer

    Quality function deployment is an example of a facilitated workshop. It starts by collecting customer needs, which also are known as the voice of the customer. The next step is to objectively sort and prioritize these needs and set goals to achieve them. Facilitated workshops are a tool and technique in the Collect Requirements process; other examples are joint application development sessions and user stories. They are all focused sessions to bring stakeholders together to define requirements, with an emphasis on those that are cross-cutting and on reconciling stakeholder differences early in the project. Since these sessions are interactive, if they are conducted effectively, they can build trust, foster relationships, and improve communications among those stakeholders who participate. An objective is to increase stakeholder consensus and to discover issues earlier and resolve them quicker than would be possible if individual sessions were used. [Planning]

    PMI®, *PMBOK® Guide*, 2013, 114
    PMI® *PMP Examination Content Outline,* 2015, Planning, 6, Task 13

16. d. Project organizational units

The WBS includes all work needed to be done to complete the project. The WBS structure can be represented by phases of the project life cycle, which are shown at the second level of decomposition. Product and project deliverables are at the third level (see Figure 5-11 in the *PMBOK® Guide*, 2013 as an example). Another way to represent the WBS is to show major deliverables at the second level of decomposition (see Figure 5-13 in the *PMBOK® Guide*, 2013). The third way is to incorporate subcomponents, which may be developed from outside of the project team such as contracted work. In this case, the seller develops the supporting contract WBS as part of the contacted work. The organizational breakdown structure (OBS) is a hierarchical representation of the project organization and shows the relationship between project activities and the organizational units responsible for these activities. [Planning]

PMI®, *PMBOK® Guide*, 2013, 126, 129–130, 548
PMI® *PMP Examination Content Outline,* 2015, Planning, 6, Task 2

17. b. Scope creep

Project scope creep is typically the result of uncontrolled changes in product or project scope without adjustments to time, cost, or resources. Scope control works to control the impact of any project scope changes, manage actual changes when they occur, and is integrated with other control processes. Scope control ensures all requested changes and recommended, corrective, or preventive actions are processed through the Perform Integrated Change Control process. [Monitoring and Controlling]

PMI®, *PMBOK® Guide*, 2013, 137
PMI® *PMP Examination Content Outline,* 2015, Monitoring and Controlling, 9, Task 2

18. c. Sum costs, schedule, and resource information

The key document generated from the create WBS process is the actual WBS. Each WBS work package is assigned to a control account with a unique identifier established for the work package from the code of accounts. These identifiers provide a structure for hierarchical summation of costs, schedule, and resource information. [Planning]

PMI®, *PMBOK® Guide*, 2013, 132
PMI® *PMP Examination Content Outline,* 2015, Planning, 6, Task 2

19. d. Project performance measurements

    Variance analysis is a tool and technique for Control Scope to determine the cause and degree of difference between the baseline and actual performance. Project performance measurements are used to assess the magnitude of variance from the original scope baseline. In scope control it is important to determine the cause and degree of the variance relevant to the scope baseline and to decide whether corrective or preventive action is required. [Monitoring and Controlling]

    PMI®, *PMBOK® Guide*, 2013, 139
    PMI® *PMP Examination Content Outline,* 2015, Monitoring and Controlling, 9, Task 1

20. b. Nominal group technique

    The nominal group technique enhances brainstorming with a voting process, which is used to rank the most useful ideas for further brainstorming or for prioritization. It is a group creativity technique used in Collect Requirements; other group creativity techniques are brainstorming, idea/mind mapping, affinity diagrams, and multiple criteria decision analysis. [Planning]

    PMI®, *PMBOK® Guide*, 2013, 115
    PMI® *PMP Examination Content Outline,* 2015, Planning, 6, Task 13

21. b. Identify and analyze the deliverables and related work

    Identifying and analyzing the deliverables and related work is the first step in the decomposition of a project, a tool and technique in the Create WBS process. Decomposition divides and subdivides the scope and deliverables into smaller parts that are more manageable. The deliverables should be defined in terms of how the project will be organized. For example, the major project deliverables may be used as the second level. Once the deliverables and related work are identified, the other steps in decomposition are to structure and organize the related work, decompose upper WBS levels into lower-level detailed components, develop and assign identification codes to the WBS components, and verify the degree of decomposition of the deliverable is appropriate. Review figure 5-11 in the *PMBOK® Guide*, 2013. [Planning]

    PMI®, *PMBOK® Guide*, 2013, 128–129
    PMI® *PMP Examination Content Outline,* 2015, Planning, 6, Task 2

22. c. Work performance information

Work performance information is an output of the Control Scope process. It includes correlated and contextualized information on how the project scope is performing against the scope baseline. It may include categories of changes received, identified scope variances and their causes, the impact on schedule or cost, and the forecast of future scope performance. It provides the foundation for making scope decisions. [Monitoring and Controlling]

PMI®, *PMBOK® Guide*, 2013, 139
PMI® *PMP Examination Content Outline,* 2015, Monitoring and Controlling, 9, Task 2

23. a. An assumption

Assumptions are factors that, for planning purposes, are considered to be true, real, or certain without proof or demonstration. They are listed in the project scope statement or in a separate log. It also is important to describe the potential impact of assumptions if they prove to be false. It is typical for project teams to identify, document, and validate assumptions as part of planning. [Planning]

PMI®, *PMBOK® Guide*, 2013, 124, 529
PMI® *PMP Examination Content Outline,* 2015, Planning, 6, Task 2

24. d. Differs from perform quality control in that validate scope is concerned with the acceptance—not the correctness—of the deliverables

Documentation that the customer has accepted completed deliverables is an output of validate scope. Acceptance criteria are formally signed off and approved by the customer or sponsor. The formal documentation also is received from the customer or sponsor acknowledging formal stakeholder acceptance of the deliverables of the project. [Monitoring and Controlling]

PMI®, *PMBOK® Guide*, 2013, 134
PMI® *PMP Examination Content Outline,* 2015, Monitoring and Controlling, 9, Task 1

25. b. Corrective action

    Recommended corrective action is an output from Control Scope. In addition to bringing expected future performance in line with the project management plan, it also serves to bring expected future performance in line with the project scope statement. [Monitoring and Controlling]

    PMI®, *PMBOK® Guide*, 2013, 93, 140, 534
    PMI® *PMP Examination Content Outline,* 2015, Monitoring and Controlling, 9, Task 2

26. a. Requirements traceability matrix

    It is an input to the Control Scope process as it helps to detect the impact of any change or deviation from the scope baseline to the project objectives. The requirements traceability matrix is a grid that links product requirements from their origin to the deliverables that satisfy them. It helps ensure each requirement adds business value by linking it to business and project objectives. See Figure 5-6 in the *PMBOK® Guide*, 2013 for an example. [Monitoring and Controlling]

    PMI®, *PMBOK® Guide*, 2013, 118–119, 138
    PMI® *PMP Examination Content Outline,* 2015, Monitoring and Controlling, 9, Task 2

27. c. Work authorization system

    The work authorization system is not used in Control Scope. The others are examples of organizational process assets that may require update as a result of scope control. Other outputs of this process are work performance information, change requests, updates to the project management plan in terms of updates to the scope baseline and other baselines, and updates to project documents, which are the requirements documentation and the requirements traceability matrix. [Monitoring and Controlling]

    PMI®, *PMBOK® Guide*, 2013, 139–140
    PMI® *PMP Examination Content Outline,* 2015, Monitoring and Controlling, 9, Task 2

28. b. Costs authorized and incurred

Work performance information is an output of Validate Scope. It emphasizes deliverables—whether or not they have started, their progress, and ones that have finished or have been accepted. This information is communicated to stakeholders. [Monitoring and Controlling]

PMI®, *PMBOK® Guide*, 2013, 136
PMI® *PMP Examination Content Outline,* 2015, Monitoring and Controlling, 9, Task 1

29. a. Phase-to-phase relationship

The requirements management plan defines how requirements will be analyzed, documented, and managed. It is strongly influenced by the phase-to-phase relationship as it influences how requirements will be managed. The project manager selects the most effective relationship for the project and documents it in the plan; many of the requirements management components are based on that relationship. [Planning]

PMI®, *PMBOK® Guide*, 2013, 42–44, 110
PMI® *PMP Examination Content Outline,* 2015, Planning, 6, Task 11

30. d. Project verification

The requirements traceability matrix is an output of the Collect Requirements process. It includes tracing requirements to business needs, opportunities, and objectives; project objectives; project scope: WBS deliverables; product design; product development; test strategy and scenarios; as well as high-level requirements to more detailed requirements. [Planning]

PMI®, *PMBOK® Guide*, 2013, 118–119
PMI® *PMP Examination Content Outline,* 2015, Planning, 6, Task 11

31. a. Should be reviewed according to the Perform Integrated Change Control process

A requested change is an output from the Control Scope process. Such a change should be handled according to the integrated change control process and may result in an update to the scope baseline or other components of the project management plan. Change requests can include preventive or corrective actions, defect repair, or enhancement requests. [Monitoring and Controlling]

PMI®, *PMBOK® Guide*, 2013, 140
PMI® *PMP Examination Content Outline,* 2015, Monitoring and Controlling, 9, Task 1

32. b. Change request

The change requests are not an input of the Validate Scope process but are an output. The other inputs are the: project management plan, requirements documentation, requirements traceability matrix, verified deliverables, and work performance information. [Monitoring and Controlling]

PMI®, *PMBOK® Guide*, 2013, 134–135
PMI® *PMP Examination Content Outline,* 2015, Monitoring and Controlling, 9, Task 1

33. d. Updates are needed to the scope baseline

If the approved change requests have an effect on the project scope, the scope statement, the WBS, and the WBS dictionary need to be revised and reissued to reflect the approved changes following the Perform Integrated Change Control process. [Monitoring and Controlling]

PMI®, *PMBOK® Guide*, 2013, 140
PMI® *PMP Examination Content Outline,* 2015, Monitoring and Controlling, 9, Task 2

34. c. Project requirements

Various components of requirements documentation can be used. Examples are: business requirements, stakeholder requirements, solution requirements, project requirements, transition requirements, and requirements assumptions, dependencies, and constraints. Project requirements consist of acceptance criteria and levels of service performance, safety, and compliance. [Planning]

PMI®, *PMBOK® Guide*, 2013, 117–118
PMI® *PMP Examination Content Outline,* 2015, Planning, 6, Task 11

35. b. Work performance information

Other inputs include all the answers listed except work performance information. Another input is work performance data. [Monitoring and Controlling]

PMI®, *PMBOK® Guide*, 2013, 134–135
PMI® *PMP Examination Content Outline,* 2015, Monitoring and Controlling, 9, Task 2

36. b. Develop a subproject WBS for the work package that is your company's responsibility

Work packages are items at the lowest level of the WBS. A subproject is a smaller portion of the original project when a project is subdivided into more manageable components or pieces. A subproject WBS then breaks down work packages into greater detail. The WBS is represented in this way to incorporate a subcomponent, which may be developed by organizations outside of the project team such as contractors. The seller develops the supporting contract WBS as part of the contracted work. [Planning]

PMI®, *PMBOK® Guide*, 2013, 129, 564
PMI® *PMP Examination Content Outline,* 2015, Planning, 6, Task 11

37. a. Is a critical component of the scope baseline

The project scope statement, along with the WBS and WBS dictionary, is a key input to Control Scope. They form the scope baseline, which is compared to actual results to determine if a change, corrective action, or preventive action is needed. [Monitoring and Controlling]

PMI®, *PMBOK® Guide*, 2013, 138
PMI® *PMP Examination Content Outline,* 2015, Monitoring and Controlling, 9, Task 1

38. c. Progressively elaborates characteristics

The project scope statement describes the deliverables and the work required to create them. It also provides a common understanding of the scope among stakeholders. The product scope statement is a key component as it progressively elaborates the characteristics of the product, service, or result in the project charter and requirements documentation. [Planning]

PMI®, *PMBOK® Guide*, 2013, 123
PMI® *PMP Examination Content Outline,* 2015, Planning, 6, Task 11

39. a. To depict product scope

It is a tool and technique in Collect Requirements and is an example of a scope model. The context diagram visually depicts the product scope as it shows a business system (process, equipment, or computer, etc.) and how people and other systems (actors) interact with it. The diagram shows inputs to the business system, the actor(s) providing the input, outputs from the business system, and actor(s) receiving the output. [Planning]

PMI®, *PMBOK® Guide*, 2013, 117
PMI® *PMP Examination Content Outline,* 2015, Planning, 6, Task 11

40. b. WBS dictionary

The WBS dictionary typically includes a code of accounts identifier, a description of work, assumptions and constraints, responsible organization, a list of schedule milestones, associated schedule activities, required resources, cost estimates, quality requirements, acceptance criteria, technical references, and agreement information. It provides detailed deliverable, activity, and scheduling information about each component in the WBS and is part of the project's scope baseline. [Planning]

PMI®, *PMBOK® Guide*, 2013, 132
PMI® *PMP Examination Content Outline,* 2015, Planning, 6, Task 11

Project Time Management

Study Hints

The Project Time Management questions on the PMP® certification exam focus heavily on the program evaluation and review technique (PERT), the critical path method (CPM), the precedence diagramming method (PDM), and the critical chain method; the differences between these four techniques; and the appropriate circumstances for their use. The exam tests your knowledge of how PERT/CPM networks are constructed, how schedules are computed, what the critical path is, and how networks are used to analyze and solve project scheduling, and resource allocation and leveling issues. There is a good chance that you will be presented with a network diagram that will be the subject of five or more questions. Therefore, detailed knowledge of network scheduling is essential. There also seems to be a focus on fast tracking as a method to accelerate the project schedule. You must know the advantages offered by networks over bar charts and network diagrams. You also should understand the concept of float (or slack) and how it presents challenges and opportunities to project schedulers. Additionally, you may see questions about the earned value formulas relative to scheduling such as the Schedule Variance (SV) and the Schedule Control Index (SPI).

Because a thorough understanding of networks and scheduling is required to successfully answer questions on Project Time Management, you can if you wish take a course relating to that topic. If you cannot take a course, you may want to consult the user's manual for one of the more popular desktop software project management packages. Also, please consult and consider reading PMI's *Practice Standard for Scheduling, Second Edition* (2011) as it contains a wealth of information to augment what is in the PMBOK. Typically, you will find plenty of illustrations and short, easy-to-understand scheduling exercises at the level of detail required to correctly answer the exam questions.

The *PMBOK® Guide* separates the function of Project Time Management into seven processes: plan schedule management, define activities, sequence activities, estimate activity resources, estimate activity durations, develop schedule, and control schedule. Review *PMBOK® Guide*, 2013, Figure 6-1 before taking the practice test. Know this chart thoroughly.

Following is a list of the major Project Time Management topics. Use it to help focus your study efforts on the areas most likely to appear on the exam.

Major Topics

Schedule management plan
Define activities

- Activity list
- Activity attributes
- Milestone list

Sequence activities

- PDM
- Dependencies
- Leads and lags
- Project schedule network diagrams

Estimate activity resources

- Expert judgment
- Alternative analysis
- Published estimating data
- Bottom-up estimating
- Activity resource requirements
- Resource breakdown structure
- Resource calendars

Estimate activity durations

- Expert judgment
- Analogous estimating
- Parametric estimates
- Three-point estimates
- Reserve analysis
- Group decision-making techniques
- Activity duration estimates

Develop schedule

- Schedule network analysis
- Critical path method
- Critical chain method
- Resource optimization techniques
- Leads and lags
- Crashing and fast-tracking
- Project schedule
- Schedule baseline
- Project calendars

Control schedule

- Performance reviews
- Work performance information
- Change requests
- Updates
- Project management plan
- Project documents

Practice Questions

INSTRUCTIONS: Note the most suitable answer for each multiple-choice question in the appropriate space on the answer sheet.

Use the following network diagram to answer questions 1 through 4. Activity names and duration are provided.

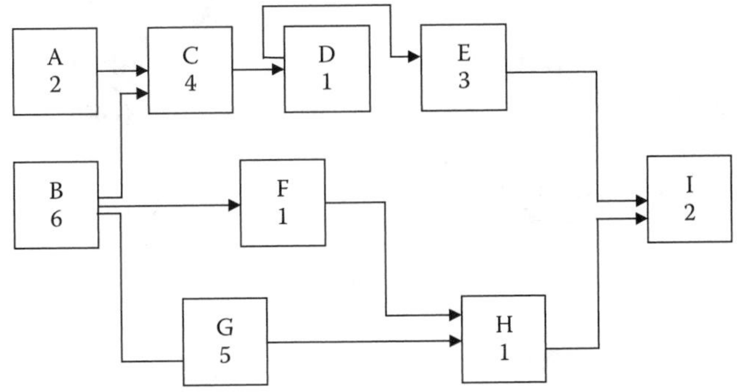

1. What is the duration of the critical path in this network?

   a. 10
   b. 12
   c. 14
   d. 15

2. What is the float for Activity G?

   a. –2
   b. 0
   c. 1
   d. 4

3. If a project planner imposes a finish time of 14 on the project with no change in the start date or activity durations, what is the total float of Activity E?

   a. –1
   b. 0
   c. 2
   d. Cannot be determined

4. If the imposed finish time in question 3 above is removed and reset to 16 and the duration of Activity H is changed to 3, what is the late finish for Activity G?

   a. –11
   b. 11
   c. –13
   d. 13

5. Your company, which operates one of the region's largest chemical processing plants, has been convicted of illegally dumping toxic substances into the local river. The court has mandated that the required cleanup activities be completed by February 15. This date is an example of—

   a. A key event
   b. A milestone
   c. A discretionary dependency
   d. An external dependency

6. You are managing a construction project for a new city water system. The contract requires you to use special titanium piping equipment that is guaranteed not to corrode. The titanium pipe must be resting in the ground a total of 10 days before connectors can be installed. In this example, the 10-day period is defined as—

   a. Lag
   b. Lead
   c. Float
   d. Slack

7. Of the following tools and techniques, which one is NOT used for Control Schedule?

   a. Fast tracking
   b. Trend analysis
   c. Three-point estimates
   d. Adjusting leads and lags

8. You are planning to conduct the team-building portion of your new project management training curriculum out-of-doors in the local park. You are limited to scheduling the course at certain times of the year, and the best time for the course to begin is mid-July. One of the more common date constraints to use as you develop the project schedule is—

    a. "Start no earlier than"
    b. "Finish no earlier than"
    c. "Fixed late start"
    d. "Fixed early finish"

9. Project schedule development is an iterative process. If the start and finish dates are not realistic, the project probably will not finish as planned. You are working with your team to define how to manage schedule changes. You documented your decisions in which of the following?

    a. Schedule change control procedures
    b. Schedule management plan
    c. Schedule risk plan
    d. Service-level agreement

10. If, when developing your project schedule, you want to define a distribution of probable results for each schedule activity and use that distribution to calculate another distribution of probable results for the total project, the most common technique to use is—

    a. What-if scenario analysis
    b. Monte Carlo analysis
    c. Linear programming
    d. Concurrent engineering

11. Your lead engineer estimates that a work package will most likely require 50 weeks to complete. It could be completed in 40 weeks if all goes well, but it could take 180 weeks in the worst case. What is the PERT estimate for the expected duration of the work package?

    a. 45 weeks
    b. 70 weeks
    c. 90 weeks
    d. 140 weeks

12. Your customer wants the project to be completed six months earlier than planned. You believe you can meet this target by overlapping project activities. The approach you plan to use is known as—

    a. Critical chain
    b. Fast tracking
    c. Leveling
    d. Crashing

13. Activity A has a duration of three days and begins on the morning of Monday the 4th. The successor activity, B, has a finish-to-start relationship with A. The finish-to-start relationship has three days of lag, and activity B has a duration of four days. Sunday is a non-workday. Such data can help to determine—

    a. The total duration of both activities is 8 days
    b. Calendar time between the start of A to the finish of B is 11 days
    c. The finish date of B is Wednesday the 13th
    d. Calendar time between the start of A to the finish of B is 14 days

14. You can use various estimating approaches to determine activity durations. When you have a limited amount of information available about your project, especially when in the early phases, the best approach to use is—

    a. Bottom-up estimating
    b. Analogous estimating
    c. Reserve analysis
    d. Parametric analysis

15. "I cannot test the software until I code the software." This expression describes which of the following dependencies?

    a. Internal
    b. Rational
    c. Preferential
    d. Mandatory or hard

16. Working with your team to provide the basis for measuring and reporting schedule progress, you agree to use the—

    a. Schedule data
    b. Network diagram
    c. Project schedule
    d. Technical baseline

17. Your project schedule was approved. Management has now mandated that the project be completed as soon as possible. However, you do not think it is possible given resource constraints. In order to convince your management of your need for additional resources, you decide to use—

    a. Resource manipulation
    b. Resource breakdown structure
    c. Critical chain scheduling
    d. Resource leveling

18. Review the following network diagram and table. Of the various activities, which ones would you crash and in what order?

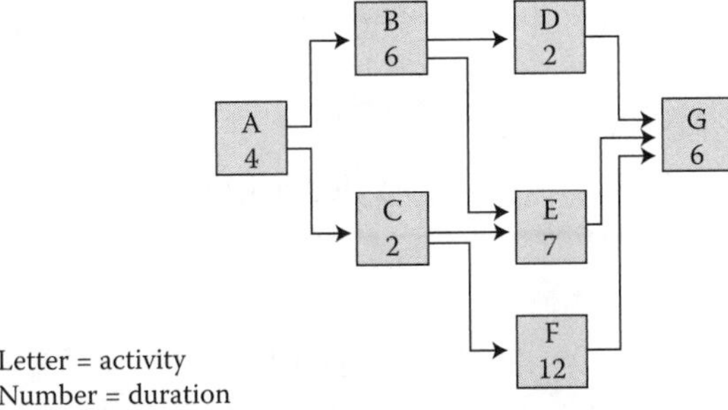

	Time required, weeks		Cost $		Crashing cost per weeks, $
Activity	Normal	Crash	Normal	Crash	
A	4	2	10,000	14,000	2000
B	6	5	30,000	42,500	12,500
C	2	1	8000	9500	1500
D	2	1	12,000	18,000	6000
E	7	5	40,000	52,000	6000
F	12	3	20,000	29,000	3000
G	6	2	5000	30,000	6000

    a. A, C, E, and F
    b. A, B, D, and F
    c. A, B, E, and F
    d. C, A, F, and G

19. You are remodeling your kitchen and decide to prepare a network diagram for this project. Your appliances must be purchased and available for installation by the time the cabinets are completed. In this example, these relationships are—

    a. Start-to-finish
    b. Finish-to-start
    c. Start-to-start
    d. Finish-to-finish

20. Decomposition is a technique used for both WBS development and activity definition. Which following statement best describes the role decomposition plays in activity definition as compared to creating the WBS?

    a. Final output is described in terms of work packages in the WBS
    b. Final output is described as deliverables or tangible items
    c. Final output is described as activities
    d. Decomposition is used the same way in scope definition and activity definition

21. When sequencing project activities in the schedule, all the following are true EXCEPT—

    a. There may be scheduled dates for specific milestones
    b. Every activity is connected to at least one predecessor and at least one successor
    c. Lead or lag time may be required
    d. Necessary sequencing of events may be described by the activity attributes

22. A schedule performance index of less than 1.0 indicates that the—

    a. Project is running behind the monetary value of the work it planned to accomplish
    b. Earned value physically accomplished thus far is 100%
    c. Project has experienced a permanent loss of time
    d. Project may not be on schedule, but the project manager need not be concerned

23. Various tools and techniques are available to sequence activities, and several factors can help to determine which tool or technique to select. When a project manager decides to include *sub-networks* or a *fragment network* as part of his or her scheduling technique, what does this decision say about the project?

    a. The work is unique requiring special network diagrams at various stages.
    b. Software that manages resources is available over an existing electronic network.
    c. Several identical or nearly identical series of activities are repeated throughout the project.
    d. Multiple critical paths exist in the project.

24. To meet regulatory requirements, you need to crash your project schedule. Your first step is to compute—
    a. The cost and time slope for each critical activity that can be expedited
    b. The cost of additional resources to be added to the project's critical path
    c. The time saved in the overall schedule when tasks are expedited on the critical path
    d. Three probabilistic time estimates of PERT for each critical path activity

25. Which one of the following is a key input to the Define Activities process?
    a. Project management plan
    b. Project scope statement
    c. Project scope baseline
    d. Project charter

26. Unlike bar charts, milestone charts show—
    a. Scheduled start or completion of major deliverables and key external interfaces
    b. Activity start and end dates of critical tasks
    c. Expected durations of the critical path
    d. Dependencies between complementary projects

27. Project managers should pay attention to critical and subcritical activities when evaluating project time performance. One way to do this is to analyze 10 subcritical paths in order of ascending float. This approach is part of—
    a. Variance analysis
    b. Simulation
    c. Earned value management
    d. Trend analysis

28. An activity has an early start date of the 10th and a late start date of the 19th. The activity has a duration of four days. There are no non-workdays. From the information given, what can be concluded about the activity?
    a. Total float for the activity is nine days.
    b. The early finish date of the activity is the end of the day on the 14th.
    c. The late finish date is the 25th.
    d. The activity can be completed in two days if the resources devoted to it are doubled.

29. In project development, schedule information, such as who will perform the work, where the work will be performed, activity type, and WBS classification, are examples of—

    a. Activity attributes
    b. Constraints
    c. Data in the WBS repository
    d. Refinements

30. Which of the following is a key benefit of Control Schedule?

    a. It updates schedule progress
    b. It minimizes risk
    c. It reprioritizes remaining work as required
    d. It describes how the schedule will be managed and controlled

31. It is important to use the critical path analysis in Control Schedule because—

    a. It assists in reviewing scenarios to bring the schedule in line with the plan
    b. It enables a consideration of the resource availability and the project time
    c. It examines project performance over time
    d. It can help identify schedule risks

32. Several types of float are found in project networks. Float that is used by a particular activity and does NOT affect the float in later activities is called—

    a. Extra float
    b. Free float
    c. Total float
    d. Expected float

33. All the following statements regarding critical chain method are true EXCEPT—

    a. It modifies the schedule to account for limited resources
    b. The first step is to use conservative estimates for activity durations
    c. Duration buffers are added on the critical path
    d. It focuses on managing buffer activity durations

34. You are managing a new technology project designed to improve the removal of hazardous waste from your city. You are in the planning phase of this project and have prepared your network diagram. Your next step is to—

    a. Describe any unusual sequencing in the network
    b. State the number resources required to complete each activity
    c. Establish a project calendar and link it to individual resource calendars
    d. Determine which schedule compression technique is the most appropriate, because your customer requests that the project be completed as soon as possible

35. The risk register may need to be updated as an output of which following process?

    a. Define Activities
    b. Sequence Activities
    c. Estimate Activity Resources
    d. Control Schedule

36. You are managing a project that will use a virtual team with team members on three different continents. Your company is looking to use the virtual team to provide a lower cost product by using resources in countries that have a favorable exchange rate to that of your country. To assist in this process as you estimate resource requirements, it is helpful to consider—

    a. Bottom-up estimating
    b. Published estimating data
    c. Analogous estimating
    d. Reserve analysis

37. Activity A has a pessimistic *(P)* estimate of 36 days, a most likely *(ML)* estimate of 21 days, and an optimistic *(O)* estimate of 6 days. What is the probability that activity A will be completed in 16 to 26 days?

    a. 55.70 percent
    b. 68.26 percent
    c. 95.46 percent
    d. 99.73 percent

38. You are managing a project to redesign a retail store layout to improve customer throughput and efficiency. Much project work must be done on site and will require the active participation of store employees who are life-long members of a powerful union with a reputation for labor unrest. One important component of your schedule must be—

    a. A resource capabilities matrix
    b. Buffers and reserves
    c. A resource calendar
    d. A resource histogram

39. To account for uncertainty in a schedule, reserve analysis may be used. All the following are examples of contingency reserves EXCEPT—

    a. Fixed number of work periods
    b. Percent of the estimated activity duration
    c. Buffers
    d. Productivity metrics

40. The reason that the schedule performance index (SPI) is shown as a ratio is to—

    a. Enable a detailed analysis of the schedule regardless of the value of the schedule variance
    b. Distinguish between critical path and noncritical path work packages
    c. Provide the ability to show performance for a specified time period for trend analysis
    d. Measure the actual time to complete the project

Answer Sheet

1.	a	b	c	d	21.	a	b	c	d
2.	a	b	c	d	22.	a	b	c	d
3.	a	b	c	d	23.	a	b	c	d
4.	a	b	c	d	24.	a	b	c	d
5.	a	b	c	d	25.	a	b	c	d
6.	a	b	c	d	26.	a	b	c	d
7.	a	b	c	d	27.	a	b	c	d
8.	a	b	c	d	28.	a	b	c	d
9.	a	b	c	d	29.	a	b	c	d
10.	a	b	c	d	30.	a	b	c	d
11.	a	b	c	d	31.	a	b	c	d
12.	a	b	c	d	32.	a	b	c	d
13.	a	b	c	d	33.	a	b	c	d
14.	a	b	c	d	34.	a	b	c	d
15.	a	b	c	d	35.	a	b	c	d
16.	a	b	c	d	36.	a	b	c	d
17.	a	b	c	d	37.	a	b	c	d
18.	a	b	c	d	38.	a	b	c	d
19.	a	b	c	d	39.	a	b	c	d
20.	a	b	c	d	40.	a	b	c	d

Answer Key

1. d. 15

   The total duration for the path B-C-D-E-I is 15. The duration of any other path in the network is less than 15. You calculate the critical path in this question through the critical path method. You are doing so to estimate the minimum project duration and the amount of schedule flexibility in the network paths. To do so, determine the early start, early finish, late start, and late finish for all the activities by performing a forward and backward pass. The critical path represents the longest path in the network, which determines the shortest possible project duration. The early and late start and finish dates are not the schedule but show the time period when the activity could be executed. The critical path method calculates the amount of scheduling flexibility on logical network paths. The critical path method normally is one that has zero total float on the critical path; note the word "normally". Activities on the critical path are critical path activities. Review the example in Figure 6-18 in the *PMBOK® Guide*, 2013. [Planning]

   PMI®, *PMBOK®Guide*, 2013, 176–177
   PMI® *PMP Examination Content Outline,* 2015, Planning, 6, Task 4

2. c. 1

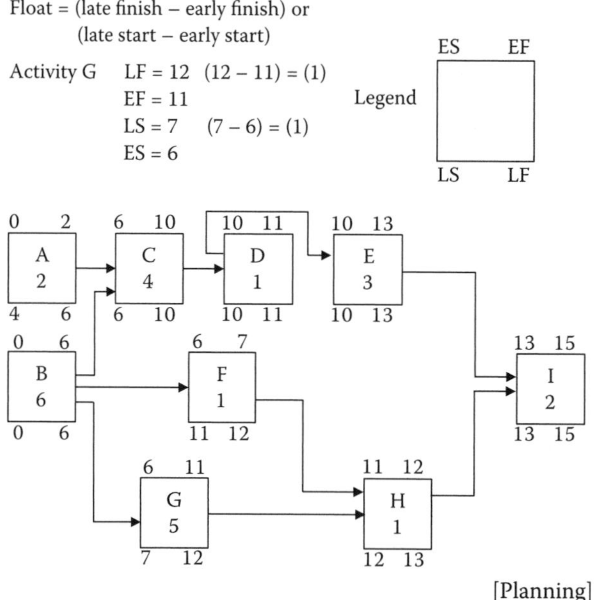

[Planning]

Float is often called slack. The critical path activities have zero float. Positive total float is caused if the backward path is calculated from a schedule constraint that is later than the early finish date calculated by the forward path. Negative total float is caused if a constraint on the late date is violated by duration and logic. Once the total float for a network path has been calculated then the free float, or the amount of time an activity can be delayed without delaying the early start of a successor or violating a schedule constraint, can be determined. Figure 6-18 in the *PMBOK® Guide*, 2013 shows the total float in the example.

PMI®, PMBOK® Guide, 2013, 177
PMI® *PMP Examination Content Outline,* 2015, Planning, 6, Task 4

3. a. –1

The imposed finish date becomes the late finish for Activity I. The late dates for each activity need to be recalculated. The dates for Activity E become—

ES = 10
EF = 13
LS = 9
LF = 12

Total float = LS – ES or 9 – 10 = (–1) or
LS – EF or 12 – 13 = (–1)

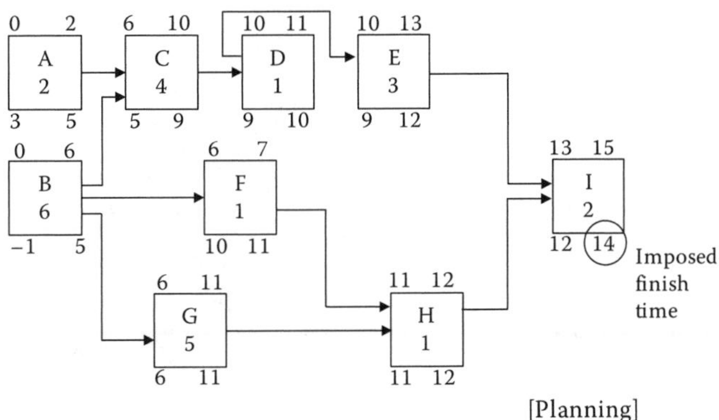

[Planning]

PMI®, *PMBOK®Guide*, 2013, 177
PMI® *PMP Examination Content Outline,* 2015, Planning, 6, Task 4

4. b. 11

The late dates for all activities need to be recalculated given the changed duration. Activity G's revised late dates are—

LF = 11
LS = 6

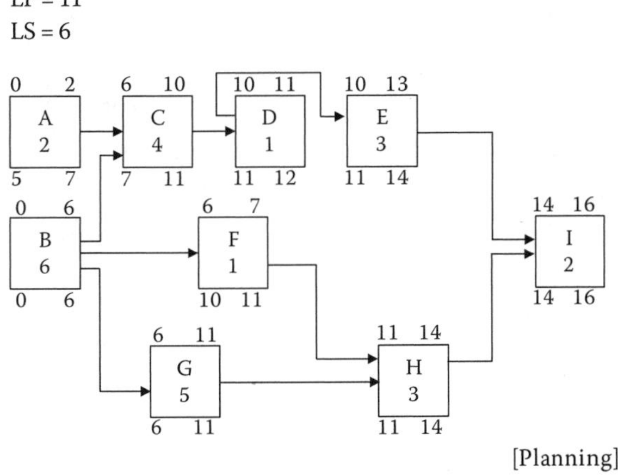

[Planning]

PMI®, *PMBOK® Guide*, 2013, 177
PMI® *PMP Examination Content Outline,* 2015, Planning, 6, Task 4

5. b. A milestone

A milestone is a significant point or event in the project. Milestones may be required by the project sponsor, customer, or other external factors, such as those required by contract, for the completion of certain deliverables. They are similar to schedule activities, with the same structure and objectives, but they have zero duration as they represent a moment in time. A milestone list is an output of Define Activities. [Planning]

PMI®, *PMBOK® Guide*, 2013, 153
PMI® *PMP Examination Content Outline,* 2015, Planning, 6, Task 4

6. a. Lag

For example, in a finish-to-start dependency with a 20-day lag, the successor activity cannot start until 20 days after the predecessor has finished. A lag is the amount of time that a successor activity will be delayed with respect to a predecessor activity. The project management team determines those dependencies that may require a lead or a lag to best define the logical relationship. Keep in mind that leads (i.e., the amount of time a successor activity can be advanced with respect to a predecessor activity) and lags should not replace schedule logic, and their use should be documented. Review Figure 6-10 in the *PMBOK® Guide*, 2013 for examples of a lead and a lag. [Planning]

PMI®, *PMBOK® Guide*, 2013, 158–159
PMI® *PMP Examination Content Outline,* 2015, Planning, 6, Task 4

7. c. Three-point estimates

Three-point estimates are used to determine the estimates that go into the schedule using the Program Evaluation and Review Technique (PERT) to better improve the accuracy of activity duration estimating. PERT calculates the most likely, optimistic, and pessimistic estimate. In Schedule Control, tools and techniques are performance reviews (including trend analysis or answer b), project management software, resource optimization techniques, modeling techniques, leads and lags (answer d), schedule compression (answer a), and a scheduling tool. [Planning and Monitoring and Controlling]

PMI®, *PMBOK® Guide*, 2013, 170–171, 188–190
PMI® *PMP Examination Content Outline,* 2015, Monitoring and Controlling, 9, Task 2
PMI® *PMP Examination Content Outline,* 2015, Planning, 5, Task 4

8. a. "Start no earlier than"

   Imposed dates on schedule activity starts of finishes can be used to restrict the start of finish to occur either no earlier than a specified date or no later than a specified date. Although all four date constraints typically are available in project management software, "start no earlier than" and "finish no later than" constraints are more commonly used. In this question given the date constraint of mid-July, start no earlier than is the best answer. [Planning]

   PMI®, *PMBOK® Guide*, 2013, 543
   PMI® *PMP Examination Content Outline,* 2015, Planning, 6, Task 4

9. b. Schedule management plan

   The schedule management plan is part of the overall project management plan. Whether it is formal, informal, highly detailed, or broadly framed, it generally is based on specific project needs. It is the output of the Plan Schedule Management process and its typical contents include: project schedule model development, level of accuracy, units of measure, organizational procedure links, project schedule model maintenance, control thresholds, rules of performance measurement, reporting formats, and process descriptions. Study these contents since this plan establishes the criteria and activities for developing, monitoring, and controlling the schedule. [Planning]

   PMI®, *PMBOK® Guide*, 2013, 148–149
   PMI® *PMP Examination Content Outline,* 2015, Planning, 6, Task 4

10. b. Monte Carlo analysis

    Simulation, a modeling technique in Develop Schedule, involves calculating multiple project durations with different sets of activity assumptions using probability distributions constructed from three-point estimates or PERT to account for uncertainty. Monte Carlo analysis is the most commonly used simulation technique. In it, a distribution of possible activity durations is defined for each activity and is used to calculate a distribution of possible outcomes for the total project. What-if scenario analysis is another modeling technique. [Planning]

    PMI®, *PMBOK® Guide*, 2013, 180
    PMI® *PMP Examination Content Outline,* 2015, Planning, 6, Task 4

11. b. 70 weeks

$$E(t) = \frac{\text{Optimistic} + (4 \times \text{Most likely}) + \text{Pessimistic}}{6}$$

$$= \frac{40 + 200 + 180}{6} = \frac{420}{6} = 70 \text{ weeks}$$

As discussed earlier, use of the three-time estimates (i.e., optimistic, most likely, and pessimistic) improves the accuracy of a single point estimate. [Planning]

PMI®, *PMBOK® Guide*, 2013, 170–171
PMI® *PMP Examination Content Outline,* 2015, Planning, 6, Task 4

12. b. Fast tracking

Fast tracking is a way to accelerate the project schedule. It is a schedule compression tool in the Develop Schedule process in which activities or phases normally done in sequence are performed in parallel for part of their duration. However, it may result in rework and increased risk. Consider it for activities that can be overlapped to shorten the project schedule. Crashing is another compression technique. In the Control Schedule process, it is a tool and technique used to find ways to bring project activities that are behind into alignment with the schedule management plan. [Planning and Monitoring and Controlling]

PMI®, *PMBOK® Guide*, 2013, 181, 190, 540
PMI® *PMP Examination Content Outline,* 2015, Planning, 6, Task 4
PMI® *PMP Examination Content Outline,* 2015, Monitoring and Controlling, 9, Task 1

13. b. Calendar time between the start of A to the finish of B is 11 days

The duration of A, which is three, is added to the duration of B, which is four, for a total of seven. The three days between the activities is lag and not duration. The lag is a constraint and must be taken into account as part of the network calculations, but it does not consume resources. The total time by the calendar is 11 days as counted from the morning of Monday the 4th. The lag occurs over Thursday, Friday, and Saturday. Sunday is a non-work day, so activity B does not start until Monday the 11th. Therefore, the calendar time is 11 days, and activity B ends on Thursday the 14th. In the Sequence Activities and Develop Schedule processes, leads and lags are a tool and technique. [Planning]

PMI®, *PMBOK® Guide*, 2013, 158–159, and 180
PMI® *PMP Examination Content Outline,* 2015, Planning, 6, Task 4

14. b. Analogous estimating

Although limitations exist with all estimating approach, analogous estimating is often used when there is a limited amount of information for the project. It uses historical information from a similar activity or a project. It uses parameters from a previous but similar project such as duration, budget, size, weight, and complexity to estimate the same parameter or measure for a future project. It is considered a gross value estimating approach, which often is adjusted for known differences in project complexity It is a tool and technique in the Estimate Activity Durations process. [Planning]

PMI®, P*MBOK® Guide*, 2013, 169
PMI® *PMP Examination Content Outline,* 2015, Planning, 6, Task 4

15. d. Mandatory or hard

Mandatory dependencies may be required contractually or may be inherent in the nature of the project work. They describe a relationship in which the successor activity cannot be started because of physical constraints until the predecessor activity has been finished. For example, software cannot be tested until it has been developed (or coded). They may be called hard logic or hard dependencies; however, technical dependencies may not be mandatory. This distinction is important to remember as it could be a possible test question. The project team determines the dependencies that are mandatory as it sequences the activities; it then is part of dependency determination, a tool and technique in the Sequence Activities process. Dependency has four attributes but only two are applied at the same time: mandatory external dependencies, mandatory internal dependencies, discretionary external dependencies, or discretionary internal dependencies. They should not be confused with assigning schedule constraints in the scheduling tool. [Planning]

PMI®, *PMBOK® Guide*, 2013, 157
PMI® *PMP Examination Content Outline,* 2015, Planning, 6, Task 4

16. c. Project schedule

The approved project schedule is a key input to the Control Schedule process. It refers to the most recent version with any notes to indicate updates, completed activities, and started activities as of the indicated data date. [Monitoring and Controlling]

PMI®, *PMBOK® Guide*, 2013, 182, 187
PMI® *PMP Examination Content Outline,* 2015, Monitoring and Controlling, 9, Task 2

17. d. Resource leveling

While resource leveling will often result in a project duration that is longer than the preliminary schedule as the original critical path probably will change and increase, it can also be used to get a schedule back on track by reassigning activities from noncritical to critical path activities. It is a resource optimization technique in the Develop Schedule process. Start and finish dates are adjusted based on resource constraints, and the goal is to balance demand with the available supply of resources. It is used when scarce or critically required resources are only available at certain times or in limited quantities or are over allocated if the resource is assigned to two activities at the same time, or to keep resource use at a constant level. Refer to Figure 6-20 in the *PMBOK® Guide*, 2013 for an example. Resource smoothing is the other resource optimization technique. [Planning]

PMI®, *PMBOK® Guide*, 2013, 179–180
PMI® *PMP Examination Content Outline,* 2015, Planning, 6, Task 4

18. d. C, A, F, and G

First, it is necessary to determine the critical path, which is A, C, F, and G. To determine the lowest weekly crashing cost, start with C at $1,500 per week. The next activity is A, followed by F and G. Understand how to crash a schedule in case you have a question on it in your exam. Along with fast tracking, it is a schedule compression technique in the Develop Schedule process. The purpose is to shorten the schedule for the least incremental cost by adding resources. Examples are to approve overtime, add additional resources, or pay to expedite delivery of activities on the critical path. It only works for activities on the critical path and does not always provide a viable alternative. The latter means it can result in increased risk and/or cost. Refer to Kerzner, 2013, pages 620-623, for a detailed example of crashing. [Planning]

PMI®, *PMBOK® Guide*, 2013, 181
Kerzner, H., *Project Management: A Systems Approach to Planning, Scheduling, and Controlling,* 2013, 620–623
PMI® *PMP Examination Content Outline,* 2015, Planning, 6, Task 4

19. d. Finish-to-finish

The completion of the work of the successor activity depends upon the completion of the work of the predecessor activity. It is a tool and technique in the precedence diagramming method in the Sequence Activities process in which there are four types of dependencies or logical relationships as listed in the answers to this question. [Planning]

PMI®, *PMBOK® Guide*, 2013, 156
PMI® *PMP Examination Content Outline,* 2015, Planning, 6, Task 4

20. c. Final output is described as activities.

In the Create WBS process, the work package is defined as the lowest level of the WBS to then estimate and manage cost and duration. In the Define Activities process, final output is described as activities. It is a tool and technique in both processes as it divides and subdivides the project scope and deliverables into smaller, more manageable parts. Activities represent the effort to create the work package in Create WBS. [Planning]

PMI®, *PMBOK® Guide*, 2013, 151
PMI® *PMP Examination Content Outline,* 2015, Planning, 6, Task 4

21. b. Every activity is connected to at least one predecessor and at least one successor

The Activity Sequence process involves identifying and documenting relationships among the project activities. However, every activity or milestone is not connected to at least one predecessor or successor except the first and last. Logical relationships are set up to create a realistic project schedule. Some leads and lags may be used. [Planning]

PMI®, *PMBOK® Guide*, 2013, 156–157
PMI® *PMP Examination Content Outline,* 2015, Planning, 6, Task 4

22. a. Project is running behind the monetary value of the work it planned to accomplish

The SPI represents how much of the originally scheduled work has been accomplished at a given period in time, thus providing the project team with insight as to whether the project is on schedule. It is a measure of schedule efficiency and is calculated by EV/PV. A SPI of 1 means the project is on schedule, and the work that has been done to date is exactly the same as the work that was planned. Other values show how much costs are over or under (as in this example). [Monitoring and Controlling]

PMI®, *PMBOK® Guide*, 2013, 189, 224
PMI® *PMP Examination Content Outline,* 2015, Monitoring and Controlling, 9, Task 1

23. c. Several identical or nearly identical series of activities are repeated throughout the project.

When identical network descriptions are repeated throughout a project, templates of those activities can be developed. If those series of tasks are repeated several times, the template can be updated several times. Software can be used with the templates to facilitate documenting and adapting them for future use. The sub-network or fragment tends to represent a sub-project or a work package and is often used to illustrate or study some potential or proposed schedule condition, such as a change in preferential schedule logic or the scope of the project. [Planning]

PMI®, *PMBOK® Guide*, 2013, 564
PMI® *PMP Examination Content Outline,* 2015, Planning, 6, Task 4

24. a. The cost and time slope for each critical activity that can be expedited

Slope = (Crash cost – Normal cost)/(Crash time – Normal time). This calculation shows the cost per day of crashing the project. The slope is negative to indicate that as the time required for a project or task decreases, the cost increases. If the costs and times are the same regardless of whether they are crashed or normal, the activity cannot be expedited. [Planning]

PMI®, *PMBOK® Guide*, 2013, 181
Kerzner, H., *Project Management: A Systems Approach to Planning, Scheduling, and Controlling,* 2013, 620–623
PMI® *PMP Examination Content Outline,* 2015, Planning, 6, Task 4

25. c. Project scope baseline

The scope baseline—made up of the scope statement, WBS, and WBS dictionary—is a key input to the Define Activities process and is used to develop the activity list that subsequently will help to create the schedule. Other inputs in this process are the schedule management plan, enterprise environmental factors (i.e., organization culture and structure, published information from commercial data bases, and the PMIS), and organizational process assets (lessons learned knowledge bases, standardized processes, templates with a standard activity list or a portion of one, and activity planning-related processes, procedures, and guidelines). [Planning]

PMI®, *PMBOK® Guide*, 2013, 151
PMI® *PMP Examination Content Outline,* 2015, Planning, 6, Task 4

26. a. Scheduled start or completion of major deliverables and key external interfaces

    Milestones are singular points in time, such as the start or completion of a significant activity or group of activities. Milestone charts are an output of the Develop Schedule process, and while they are similar to bar charts, their focus is only on the answer to this question. [Planning]

    PMI®, *PMBOK® Guide*, 2013, 182
    PMI® *PMP Examination Content Outline,* 2015, Planning, 6, Task 4

27. a. Variance analysis

    Performance of variance analysis during the Control Schedule process is a key element of schedule control. Float variance is an essential planning component for evaluating project time performance. For those projects not using earned value, variance analysis can be performed by comparing the planned activity start or finish dates to actuals to identify variances between the schedule baseline and actual project performance. Then, future analysis can be done to determine the cause and degree of variance relative to the schedule baseline and needed corrective or preventive actions. In earned value analysis, the total float and early finish variances are essential planning components to determine the cause and degree of variance relative to the schedule baseline as part of schedule control, [Monitoring and Controlling]

    PMI®, *PMBOK® Guide*, 2013, 189
    PMI® *PMP Examination Content Outline,* 2015, Monitoring and Controlling 9, Task 1

28. a. Total float for the activity is nine days.

    Total float or slack is computed by subtracting the early start date from the late start date, or 19 – 10 = 9. To compute the early finish date given a duration of 4, we would start counting the activity on the morning of the 10th; therefore, the activity would be completed at the end of day 13, not 14 (10, 11, 12, 13). If we started the activity on its late start date on the morning of the 19th, we would finish at the end of day 22, not 25. Insufficient information is provided to determine whether this activity can be completed in 2 days if the resources are doubled. [Planning]

    Meredith, J.R. and Mantel, Jr., S.J., *Project Management: A Managerial Approach*, 2012, 352–353
    PMI®, *PMBOK® Guide*, 2013, 177
    PMI® *PMP Examination Content Outline,* 2015, Planning, 6, Task 4

29. a. Activity attributes

Identifying activity attributes is helpful for further selection and sorting of planned activities. They are used for schedule development and for report formatting purposes and are an output of the Define Activity process. They are not milestones as they have durations when work of the activity is being done and may have costs based on needed resources. Therefore, they extend the description of the activity as they identify its multiple components, which evolve over time. Components in the early stages of a project include the activity identifier, the WBS identifier, and the activity name. When completed, these components may have activity codes, descriptions, predecessor and successor activities, logical relationships, leads and lags, resource requirements, imposed dates, constraints, and assumptions. They can be used to identify the person responsible for doing the work, geographic area or place where the work is done, the project calendar assigned to the activity, and the activity type (i.e., level of effort, discrete effort, and apportioned effort). [Planning]

PMI®, *PMBOK® Guide*, 2013, 153
PMI® *PMP Examination Content Outline,* 2015, Planning, 6, Task 4

30. b. It minimizes risk

In Control Schedule the key benefit is that it provides the means to recognize deviations from the schedule management plan and take as required corrective and preventive action, which minimize risk. [Monitoring and Controlling]

PMI®, *PMBOK® Guide*, 2013, 185
PMI® *PMP Examination Content Outline,* 2015, Monitoring and Controlling, 9, Task 4

31. d. It can help identify schedule risks

In Control Schedule, performance reviews are a tool and technique to measure, update, and analyze schedule performance. Various techniques, such as trend analysis, the critical path method, the critical chain method, and earned value management, are used. The critical path method compares progress along the critical path to help determine schedule status. The variance on the critical path is important as it can have a direct impact on the project's end date. Also by evaluating the progress of activities on near critical paths, one can identify schedule risk. [Monitoring and Controlling]

PMI®, *PMBOK® Guide*, 2013, 188
PMI® *PMP Examination Content Outline,* 2015, Monitoring and Controlling, 9, Task 1

32. b. Free float

    Free float is defined as the amount of time an activity can be delayed without delaying the early start of any immediately succeeding activities or violating a schedule constraint. [Planning]

    PMI®, *PMBOK® Guide*, 2013, 177
    PMI® *PMP Examination Content Outline,* 2015, Planning, 6, Task 4

33. b. The first step is to use conservative estimates for activity durations

    When using critical chain techniques, the initial project schedule is developed from the critical path method and considers the effects of resource allocation, resource optimization, resource leveling, and activity duration estimating on the critical path. The durations do not include safety margins, logical relationships, and resource availability with statistically determined buffers of the aggregated safety margins at specified points on the project schedule path to account for limited resources and uncertainties in the project. The critical chain method allows a project team to place buffers on any project schedule path considering these resource limitations and project uncertainties. Review the example on page 178 in the *PMBOK® Guide*, 2013. [Planning]

    PMI®, *PMBOK® Guide*, 2013, 178
    PMI® *PMP Examination Content Outline,* 2015, Planning, 6, Task 4

34. a. Describe any unusual sequencing in the network

    A summary narrative can accompany the schedule network diagram and describe the basic approach used to sequence the activities in the network. This narrative also should describe any unusual sequences in the network. The project schedule network diagrams are an output of the Sequence Activities process. The project schedule network diagram is a graphic representation of the logical relationships or dependencies among the project schedule activities. Review Figure 6-11 on page 160 of the *PMBOK® Guide*, 2013 as an example. [Planning]

    PMI®, *PMBOK® Guide*, 2013, 159–160
    PMI® *PMP Examination Content Outline,* 2015, Planning, 6, Task 4

35. b. Sequence Activities

The risk register may require updates in both the Sequence Activities and Develop Schedule processes. In the Sequence Activity process, the activity lists, activity attributes, and milestone list may need updates as well. In the Develop Schedule process the risk register may be updated to reflect any opportunities or threats from scheduling assumptions. [Planning]

PMI®, *PMBOK® Guide*, 2013, 160, 185
PMI® *PMP Examination Content Outline,* 2015, Planning, 4, Task 10

36. b. Published estimating data

In estimating activity resources, published estimating data is a tool and technique that is used as many companies routinely publish updated production rates and unit costs of resources. This includes labor, material, and equipment for different countries and geographic locations in these countries. [Planning]

PMI®, *PMBOK® Guide*, 2013, 164
PMI® *PMP Examination Content Outline,* 2015, Planning, 6, Task 4

37. b. 68.26 percent

First, compute the standard deviation:

$$\sigma = \frac{P - O}{6} \quad \text{or} \quad \frac{36 - 6}{6} = 5 \text{ days}$$

Next, compute PERT expected time:

$$\frac{P + 4(ML) + O}{6} \quad \text{or} \quad \frac{36 + 4(21) + 6}{6} = 21 \text{ days}$$

Finally, determine range of outcomes using 1σ:

21–5 = 16 days, and 21 + 5 = 26 days

Simply defined, 1σ is the amount on either side of the mean of a normal distribution that will contain approximately 68.26 percent of the population. Duration estimates based on PERT with an assumed distribution provide an expected duration and clarify the range of uncertainty around the expected duration. [Planning]

Meredith, J.R. and Mantel, Jr., S.J., *Project Management: A Managerial Approach*, 2012, 348–350
PMI®, *PMBOK® Guide*, 2013, 170–171
PMI® *PMP Examination Content Outline,* 2015, Planning, 6, Task 4

38. c. A resource calendar

Project and resource calendars identify periods when work is allowed. Project calendars affect all resources. Resource calendars affect a specific resource or a resource category, such as a labor contract that requires certain workers to work on certain days of the week. The resource calendar identifies the working days and shifts that each specific resource is available and is used to estimate resource use. The resource calendars specify when and how long identified project resources are available during the project, and this information may be at the activity or project level. This knowledge includes considerations of resource experience, skill level, or both, as well as, where the resources are located and when they may be available. Resource calendars are an input to Estimate Activity Resources. Project calendars are an output of Develop Schedule and identify working days and shifts available for the scheduled activities. They distinguish time periods in days or parts of days available to complete scheduled activities from time periods that are not available. Project calendars may require updates. [Planning]

PMI®, *PMBOK® Guide*, 2013, 163, 184, 558
PMI® *PMP Examination Content Outline,* 2015, Planning, 6, Task 4

39. d. Productivity metrics

Duration estimates may include contingency reserves, and contingency should be identified clearly in schedule documentation. They are built into the overall project schedule to account for uncertainty. They also may be developed using such quantitative analysis methods such as Monte Carlo simulation. When more information is known about the project, the contingency reserve may be used, reduced, or eliminated. They are part of reserve analysis, a tool and technique in the Estimate Activity Durations process. [Planning]

PMI®, *PMBOK® Guide*, 2013, 171
PMI® *PMP Examination Content Outline,* 2015, Planning, 6, Task 4

40. c. Provide the ability to show performance for a specified time period for trend analysis

Because the schedule performance index (SPI) and the cost performance index (CPI) are expressed as ratios, they can be used to show performance for a specific time period or trends over a long-time horizon. The SPI measures schedule efficiency as EV/PV and measures how efficiently the project team is using its time If it is less than 1, less work was completed than planned; greater than 1 shows more work was completed than was planned. It is important to also analyze performance on the critical path to determine if the project will finish ahead of or behind finish dates. [Monitoring and Controlling]

PMI®, *PMBOK® Guide*, 2013, 190, 219, 224
PMI® *PMP Examination Content Outline,* 2015, Monitoring and Controlling, 9, Task 1

Project Cost Management

Study Hints

You do not need to be a certified public accountant to successfully answer the Project Cost Management questions on the PMP® certification exam. PMI® addresses cost management from a project manager's perspective, which is much more general than that of an accountant. However, these questions are not easy. Far from it! Exam takers find the Project Cost Management questions more difficult than most of the others because they address such a broad range of cost issues (for example, cost estimating, earned value, and creating and interpreting S-curves) and require a significant amount of study time.

You may find questions relating to contract cost management. Because cost considerations are heavily affected by contract type, and Project Procurement Management is one of the ten *PMBOK® Guide* areas on which you will be tested, time spent studying that area will help to prepare you for the cost questions on the exam and vice versa.

The exam may include several questions that require you to know and solve specific, albeit simple, formulas. You *must* have a thorough knowledge of earned value—what it is and how it is computed. Study Table 7-1 in the *PMBOK® Guide*, 2013 as it provides information on the formulas, how to calculate them, and how to interpret the results.

For additional information on project estimating, you may wish to review PMI®'s *Practice Standard for Project Estimating*, 2011. For additional information on earned value analysis from a PMI® perspective, you may also wish to consult PMI®'s *Practice Standard for Earned Value Management*, Second Edition, 2011.

PMI® views Project Cost Management as a four-step process comprising Plan Cost Management, Estimate Costs, Determine Budges, and Control Costs. See *PMBOK® Guide, 2013* Figure 7-1 for an overview of this structure. Know this chart thoroughly.

Important: PMI® allows the use of standard six-function (+, −, ×, ÷, √,%) business calculators. These calculators are provided at the exam site for your use during the exam.

Following is a list of the major Project Cost Management topics. Use it to help focus your study efforts on the areas most likely to appear on the exam.

Major Topics

Project cost management
Life-cycle cost (LCC)
Cost management plan
Estimate costs

- Scope baseline
- Human resource plan
- Project schedule
- Risk register
- Cost estimating methods
- Analogous estimating
- Parametric modeling
- Bottom-up estimating
- Three-point estimates
- Vendor bid analysis
- Reserve analysis
- Cost of quality
- Accuracy of estimates
- Order of magnitude
- Budget
- Definitive
- Direct versus indirect costs
- Contingency/management reserve
- Activity cost estimates
- Basis of estimates

Cost risk and contract type
Determine budgets

- Cost aggregation
- Reserve analysis
- Funding limit reconciliation

Cost baseline
Control costs

- Performance reviews
- Variance analysis
- Forecasting

Earned value management (EVM)
The most rudimentary building blocks

- Cost variance (CV)
- Schedule variance (SV)
- Cost performance index (CPI)
- Forecasting
- Schedule performance index (SPI)
- Budget at completion (BAC)
- Variance at completion (VAC)
- Estimate to complete (ETC)
- Estimate at completion (EAC)
- To-complete performance index (TCPI)

Earned value measurement techniques

- Weighted milestones
- Fixed formula
- Percent complete

Work performance measurements

Practice Questions

INSTRUCTIONS: Note the most suitable answer for each multiple-choice question in the appropriate space on the answer sheet.

You are using earned value progress reporting for your current project in an effort to teach your software developers the benefits of earned value. You plan to display project results on the cafeteria bulletin board so that the team knows how the project is progressing. Use the current status, listed below, to answer questions 1 through 4:

PV = $2,200
EV = $2,000
AC = $2,500
BAC = $10,000

1. According to earned value analysis, the SV and status of the project described above is—

   a. −$300; the project is ahead of schedule
   b. +$8,000; the project is on schedule
   c. +$200; the project is ahead of schedule
   d. −$200; the project is behind schedule

2. What is the CPI for this project, and what does it tell us about cost performance thus far?

   a. 0.20; actual costs are exactly as planned
   b. 0.80; actual costs have exceeded planned costs
   c. 0.80; actual costs are less than planned costs
   d. 1.25; actual costs have exceeded planned costs

3. The CV for this project is—

   a. 300
   b. −$300
   c. 500
   d. −$500

4. What is the EAC for this project, and what does it represent?

   a. $12,500; the revised estimate for total project cost (based on performance thus far)
   b. $10,000; the revised estimate for total project cost (based on performance thus far)
   c. $12,500; the original project budget
   d. $10,000; the original project budget

5. You have now prepared your cost management plan so now you are preparing your project's cost estimate. You decided to use analogous estimating. Which of the following is NOT a characteristic of analogous estimating?

   a. Supports top-down estimating
   b. Is a form of expert judgment
   c. Has an accuracy rate of ±10% of actual costs
   d. Involves using the cost of a previous, similar project as the basis for estimating current project cost

6. All the following are outputs of the Estimate Costs process EXCEPT—

   a. Activity cost estimates
   b. Basis of estimates
   c. Documented constraints
   d. Cost baseline

7. You must consider direct costs, indirect costs, overhead costs, and general and administrative costs during cost estimating. Which of the following is NOT an example of a direct cost?

   a. Salary of the project manager
   b. Subcontractor expenses
   c. Materials used by the project
   d. Electricity

8. If the cost variance is the same as the schedule variance and both numbers are greater than zero, then—

   a. The cost variance is due to the schedule variance
   b. The variance is favorable to the project
   c. The schedule variance can be easily corrected
   d. Labor rates have escalated since the project began

9. You are responsible for preparing a cost estimate for a large World Bank project. You decide to prepare a bottom-up estimate because your estimate needs to be as accurate as possible. Your first step is to—

   a. Locate a computerized tool to assist in the process
   b. Use the cost estimate from a previous project to help you prepare this estimate
   c. Identify and estimate the cost for each work package or activity
   d. Consult with subject matter experts and use their suggestions as the basis for your estimate

10. Management has grown weary of the many surprises, mostly negative, that occur on your projects. In an effort to provide stakeholders with an effective performance metric, you will use the to-complete performance index (TCPI). Its purpose is to—
    a. Determine the schedule and cost performance needed to complete the remaining work within management's financial goal for the project
    b. Determine the cost performance needed to complete the remaining work within management's financial goal for the project
    c. Predict final project costs
    d. Predict final project schedule and costs

11. If operations on a work package were estimated to cost $1,500 and finish today but, instead, have cost $1,350 and are only two-thirds complete, the cost variance is—
    a. $150
    b. –$150
    c. –$350
    d. –$500

12. When you review cost performance data on your project, different responses will be required depending on the degree of variance or control thresholds from the baseline. For example, a variance of 10 percent might not require immediate action, whereas a variance of 100 percent will require investigation. A description of how you plan to manage cost variances should be included in the—
    a. Cost management plan
    b. Change management plan
    c. Performance measurement plan
    d. Variance management plan

13. As of the fourth month on the Acme project, cumulative planned expenditures were $100,000. Actual expenditures totaled $120,000. How is the Acme project doing?
    a. It is ahead of schedule.
    b. It is in trouble because of a cost overrun.
    c. It will finish within the original budget.
    d. The information is insufficient to make an assessment.

14. On your project, you need to assign costs to the time period in which they are incurred. To do this, you should—

    a. Identify the project components so that costs can be allocated
    b. Use the project schedule as an input to determine budget
    c. Prepare a detailed and accurate cost estimate
    d. Prepare a cost performance plan

15. You have a number of costs to track and manage because your project is technically very complex. They include direct costs and indirect (overhead) costs. You have found that managing overhead costs is particularly difficult because they—

    a. Are handled on a project-by-project basis
    b. Represent only direct labor costs
    c. Represent only equipment and materials needed for the project
    d. Are usually beyond the project manager's control

16. If you want to calculate the ETC based on the assumption that work is proceeding as planned, the remaining work can be calculated by which of the following formulas?

    a. ETC = BAC – EV
    b. ETC = (BAC – EV)/CPI
    c. ETC = AC + EAC
    d. ETC = EAC – AC

17. You receive a frantic phone call from your vice president who says she is going to meet with a prospective client in 15 minutes to discuss a large and complex project. She asks you how much the project will cost. You quickly think of some similar past projects, factor in a few unknowns, and give her a number. What type of estimate did you just provide?

    a. Definitive
    b. Budget
    c. Order-of-magnitude
    d. Detailed

18. Your approved cost baseline has changed because of a major scope change on your project. Your next step should be to—

    a. Estimate the magnitude of the scope change
    b. Issue a change request
    c. Document lessons learned
    d. Execute the approved scope change

19. Which of the following is a tool for analyzing a design, determining its functions, and assessing how to provide those functions' cost effectively?

    a. Pareto diagram
    b. Value analysis
    c. Configuration management
    d. Value engineering

20. There are many useful EVM metrics, but the most critical is—

    a. CPI
    b. EAC
    c. TCPI
    d. VAC

21. The approved, integrated scope-schedule-cost plan for project work is—

    a. Project budget baseline
    b. Performance measurement baseline
    c. Level-of-effort control accounts
    d. Expressed in the To Complete Performance Index

22. It is expensive to lease office space in cities around the world. Office space can cost approximately USD $80 per square foot in Tampa, Florida. And it can cost approximately ¥50,000 per square meter in Tokyo. These "averages" can help a person to determine how much it will cost to lease office space in these cities based on the amount of space leased. These estimates are examples of—

    a. Variance analysis
    b. Parametric estimating
    c. Bottom-up estimating
    d. Reserve analysis

23. Your project manager has requested that you provide him with a forecast of project costs for the next 12 months. He needs this information to determine if the budget should be increased or decreased on this major construction project. In addition to the usual information sources, which of the following should you also consider?

    a. Cost estimates from similar projects
    b. WBS
    c. Project schedule
    d. Costs that have been authorized and incurred

24. There are a number of different earned value management rules of performance measurement that can be established as part of the cost management plan. Which one of the following is NOT an example of such a rule?

    a. Code of accounts allocation provision
    b. Formulas to determine the EAC
    c. Earned value measurement techniques
    d. Definition of the WBS level

25. Which of the following calculations CANNOT be used to determine EAC?

    a. EAC = ETC-AC
    b. EAC = BAC/CPI
    c. EAC = AC + BAC—EV
    d. EAC = AC + (BAC-EV)/(CPI x SPI)

26. Typically, the statement "no one likes to estimate, because they know their estimate will be proven incorrect" is true. However, you have been given the challenge of estimating the costs for your nuclear reactor project. A basic assumption that you need to make early in this process is—

    a. How direct and indirect costs will be handled
    b. Whether or not experts will be available to assist you in this process
    c. If there will be a multiyear project budget
    d. Whether the project has required delivery dates

27. By reviewing cumulative cost curves, the project manager can monitor—

    a. EAC
    b. PV, EV, and AC
    c. CVs and SVs
    d. CPI and SPI

28. Control accounts—

    a. Are charge accounts for personnel time management
    b. Summarize project costs at level 2 of the WBS
    c. Identify and track management reserves
    d. Represent the basic level at which project performance is measured and reported

29. Performance review meetings are held to assess schedule activity and work packages over-running or under-running the budget and to determine any estimated funds needed to complete work in progress. Typically, if EV is being used, all but which of the following information is determined?

    a. Variance analysis
    b. Trend analysis
    c. Time reporting systems
    d. Earned value performance

30. Overall cost estimates must be allocated to individual activities to establish the cost performance baseline. In an ideal situation, a project manager would prefer to prepare estimates—

    a. Before the budget is complete
    b. After the budget is approved by management
    c. Using a parametric estimating technique and model specific for that project type
    d. Using a bottom-up estimating technique

31. One way to engage team members to improve estimate accuracy is to—

    a. Hold a focus group
    b. Use vendor bid analysis
    c. Use a Delphi approach
    d. Involve industry groups

32. Assume you have used reserve analysis as a tool and technique to Control Costs. However, during this project you have focused on continual risk identification and subsequent analysis of the identified risks. In fact you had a risk expert on your team. This means you may need to—

    a. Use management reserve
    b. Re-baseline your cost estimate
    c. Perform an assumptions analysis
    d. Request additional contingency reserves for your budget

33. Increased attention to return on investment (ROI) now requires you to re-estimate the costs for your project. When you looked at how costs were first estimated, you realized an order-of-magnitude estimate was prepared, which was never refined. Therefore, now you are estimating costs from the beginning. This project is the development of a new smart TV and automated home, and you realize market conditions should be reviewed because—

    a. A competitor may be working on the same product
    b. Global supply and demand conditions can influence resource costs
    c. Resource cost data are available in your company's knowledge transfer system
    d. Expert judgment can assist in determining when costs exceed profit

34. A revised cost baseline may be required in cost control when—

    a. CVs are severe, and a realistic measure of performance is needed
    b. Updated cost estimates are prepared and distributed to stakeholders
    c. Corrective action must be taken to bring expected future performance in line with the project plan
    d. EAC shows that additional funds are needed to complete the project even if a scope change is not needed

35. As project manager, you identified a number of acceptable tolerances as part of your earned value management system. During execution, some "unacceptable" variances occurred. After each "unacceptable" variance occurred, a best practice is to—

    a. Update the budget
    b. Prepare a revised cost estimate
    c. Adjust the project plan
    d. Document lessons learned

36. Assume that the project cost estimates have been prepared for each activity and the basis of these estimates has been determined. Now, as the project manager for your nutrition awareness program in your hospital, you are preparing your budget. Because you have estimates for more than 1,200 separate activities, you have decided to first—

    a. Aggregate these estimates by work packages
    b. Aggregate these estimates by control accounts to facilitate the use of earned value management
    c. Use the results of previous projects to predict total costs
    d. Set your cost performance baseline

37. Assume you are using earned value on your project, and as of today, your cumulative CPI is below the baseline. This means—

    a. All future work will need to be performed immediately with the range of the TCPI.
    b. The ETC work will be performed at an efficiency rate that considers both the schedule and performance indices.
    c. What the project has experienced to date is expected to continue
    d. The budget should be re-baselined

38. Assume it has become obvious on your project that the budget at completion is no longer viable. This means as the project manager, you should—

    a. Recommend to your managers that the project be terminated
    b. Perform an analysis of the remaining tasks on the critical path and take corrective or preventive action as required
    c. Determine the relationship of physical performance to costs spent
    d. Consider the forecasted EAC

39. Assume that your actual costs are $800; your planned value is $1,200; and your earned value is $1,000. Based on these data, what can be determined regarding your schedule variance?

    a. At +$200, the situation is favorable as physical progress is being accomplished ahead of your plan.
    b. At –$200, the physical progress is being accomplished at a slower rate than is planned, indicating an unfavorable situation.
    c. At +$400, the situation is favorable as physical progress is being accomplished at a lower cost than was forecasted.
    d. At –$200, you have a behind-schedule condition, and your critical path has slipped.

40. The CPI on your project is 0.84. This means that you should—

    a. Place emphasis on improving the timeliness of the physical progress
    b. Reassess the life-cycle costs of your product, including the length of the life-cycle phase
    c. Recognize that your original estimates were fundamentally flawed, and your project is in an atypical situation
    d. Place emphasis on improving the productivity by which work was being performed

Answer Sheet

1.	a	b	c	d	21.	a	b	c	d
2.	a	b	c	d	22.	a	b	c	d
3.	a	b	c	d	23.	a	b	c	d
4.	a	b	c	d	24.	a	b	c	d
5.	a	b	c	d	25.	a	b	c	d
6.	a	b	c	d	26.	a	b	c	d
7.	a	b	c	d	27.	a	b	c	d
8.	a	b	c	d	28.	a	b	c	d
9.	a	b	c	d	29.	a	b	c	d
10.	a	b	c	d	30.	a	b	c	d
11.	a	b	c	d	31.	a	b	c	d
12.	a	b	c	d	32.	a	b	c	d
13.	a	b	c	d	33.	a	b	c	d
14.	a	b	c	d	34.	a	b	c	d
15.	a	b	c	d	35.	a	b	c	d
16.	a	b	c	d	36.	a	b	c	d
17.	a	b	c	d	37.	a	b	c	d
18.	a	b	c	d	38.	a	b	c	d
19.	a	b	c	d	39.	a	b	c	d
20.	a	b	c	d	40.	a	b	c	d

Answer Key

1. d. –$200; the project is behind schedule

   SV is calculated as EV – PV (in this case, $2,000 – $2,200). It is the amount by which the project is ahead or behind the planned delivery date at a given point of time. If it is positive, it shows the project is ahead of schedule; if it is neutral, it is on schedule. In this example, there is a negative variance means that the work completed is less than what was planned for at that point in the project, and it is behind schedule. [Monitoring and Controlling]

   PMI®, *PMBOK® Guide*, 2013, 224
   PMI® *PMP Examination Content Outline,* 2015, Monitoring and Controlling, 9, Task 1

2. b. 0.80; actual costs have exceeded planned costs

   CPI is calculated as EV/AC (in this case, $2,000/$2,500). EV measures the budgeted dollar value of the work that has actually been accomplished, whereas AC measures the actual cost of getting that work done. If the two numbers are the same, work on the project is being accomplished for exactly the budgeted amount of money (and the ratio will be equal to 1.0). If actual costs exceed budgeted costs (as in this example), AC will be larger than EV, and the ratio will be less than 1.0. CPI is also an index of efficiency. In this example, an index of 0.80 (or 80 percent) means that for every dollar spent on the project only 80 cents worth of work is actually accomplished. At this time, the project is over the planned costs. [Monitoring and Controlling]

   PMI®, *PMBOK® Guide*, 2013, 224
   PMI® *PMP Examination Content Outline,* 2015, Monitoring and Controlling, 9, Task 1

3. d. –$500

   CV is calculated as EV – AC (in this case, $2,000 – $2,500). A negative CV means that accomplishing work on the project is costing more than was budgeted. CV measures the amount of budget surplus or deficit at a point in time, and if it is positive, it is under the planned costs; neutral means it is on the planned costs. [Monitoring and Controlling]

   PMI®, *PMBOK® Guide*, 2013, 224
   PMI® *PMP Examination Content Outline,* 2015, Monitoring and Controlling, 9, Task 1

4. a. $12,500; the revised estimate for total project cost (based on performance thus far)

EAC is calculated as BAC/CPI (in this case, $10,000/0.80). It is now known that the project will cost more than the original estimate of $10,000. The project has been getting only 80 cents worth of work done for every dollar spent (CPI), and this information has been used to forecast total project costs. This approach assumes that performance for the remainder of the project will also be based on a CPI of 0.80. The EAC is the expected total cost of completing all work expressed as the sum of the actual cost and the estimate to complete (ETC). Recognize there are four ways to calculate the EAC based on different assumptions and learn when to use each method. [Monitoring and Controlling]

PMI®, *PMBOK® Guide*, 2013, 224
PMI® *PMP Examination Content Outline,* 2015, Monitoring and Controlling, 9, Task 1

5. c. Has an accuracy rate of ±10% of actual costs

A frequently used method of estimate costs, the analogous technique relies on experience and knowledge gained to predict future events. This technique provides planners with some idea of the magnitude of project costs but generally not within ±10%. It is considered to be a gross value estimating technique, and it is generally less costly and less time consuming than other techniques but also is generally less accurate. It is more reliable when the previous projects are similar, and when the project team members doing the estimating have the needed expertise. [Planning]

PMI®, *PMBOK® Guide*, 2013, 204–205
PMI® *PMP Examination Content Outline,* 2015, Planning, 6, Task 3

6. d. Cost baseline

The cost baseline is an output from the Determine Budget process. It is the approved version of the time-phased budget excluding any management reserves and is used for comparison to actual costs. It is developed by summarizing approved budgets for different schedule activities. Review Figure 7-8 in the *PMBOK® Guide*, 2013 to see the components of the project budget and Figure 7-9 to see the cost baseline, expenditures, and funding requirements. [Planning]

PMI®, *PMBOK® Guide*, 2013, 212–214
PMI® *PMP Examination Content Outline,* 2015, Planning, 6, Task 3

7. d. Electricity

   Direct costs are incurred for the exclusive benefit of a project (for example, salary of the project manager, materials used by the project, and subcontractor expenses). Indirect costs, also called overhead costs, are allocated to a project by its performing organization as a cost of doing business. These costs cannot be traced to a specific project and are accumulated and allocated equitably over multiple projects (for example, security guards, fringe benefits, and electricity). Costs are estimated for all resources to be charged to the project. [Planning]

   PMI®, *PMBOK® Guide*, 2013, 202
   PMI® *PMP Examination Content Outline,* 2015, Planning, 6, Task 3

8. b. The variance is favorable to the project

   A positive schedule variance indicates that the project is ahead of schedule. A positive cost variance indicates that the project has incurred less cost than estimated for the work accomplished; therefore, the project is under budget and ahead of schedule. The CV is calculated as EV-AC, and the SV is calculated as the EV-PV. [Monitoring and Controlling]

   PMI®, *PMBOK® Guide*, 2013, 224
   PMI® *PMP Examination Content Outline,* 2015, Monitoring and Controlling, 9, Task 1

9. c. Identify and estimate the cost for each work package or activity

   Bottom-up estimating is a method of estimating a component of work. It is derived by first estimating the cost of the individual work packages or activities to the greatest level of specified detail. Then this detailed cost is summarized or 'rolled-up' to higher levels for reporting and tracking. The cost and accuracy are influenced by the size and complexity of the activity or the work package. It is a tool and technique in the Estimate Costs process. [Planning]

   PMI®, *PMBOK® Guide*, 2013, 205
   PMI® *PMP Examination Content Outline,* 2015, Planning, 6, Task 3

10. b. Determine the cost performance needed to complete the remaining work within management's financial goal for the project

The TCPI takes the value of work remaining and divides it by the value of funds remaining to obtain the cost performance factor needed to complete all remaining work according to a financial goal set by management. There are two methods to calculate the TCPI, and you should be familiar with each one and when it should be used. In this example, the TCPI is calculated by (BAC-EV)/(BAC-AC). If it is greater than one, it is harder to complete; if it is 1, it is the same to complete, and less than one means it is easier to complete. It thus measures the cost performance that must be achieved with the remaining resources to meet a specified management goal and is expressed as a ratio of the cost to finish the outstanding work to the budget that is available. [Monitoring and Controlling]

PMI®, *PMBOK® Guide*, 2013, 224
PMI® *PMP Examination Content Outline,* 2015, Monitoring and Controlling, 9, Task 1

11. c. –$350

CV is calculated by EV – AC, or $1,500(2/3) – $1,350 = –$350. It is used to determine the amount of budget deficit (as in this example) or surplus at any given point in time. In this example, the project is over the planned cost; a positive value indicates it is under the planned costs, while a neutral value shows it on the planned costs. [Monitoring and Controlling]

PMI®, *PMBOK® Guide*, 2013, 224
PMI® *PMP Examination Content Outline,* 2015, Monitoring and Controlling, 9, Task 1

12. a. Cost management plan

The management and control of costs focuses on variance thresholds. Certain variances are acceptable, and others, usually those falling outside a particular range, are unacceptable. They are typically expressed as percentage deviations from the baseline plan. The actions taken by the project manager for variances are described in the cost management plan. It is the output of the Plan Cost Management Plan process and also can include: units of measure, level of precision, level of accuracy, links to organizational procedures such as the control account in the WBS, and rules of performance measurement such as earned value management. It describes how project costs will be planned, structured, and controlled. [Planning]

PMI®, *PMBOK® Guide*, 2013, 199–200
PMI® *PMP Examination Content Outline,* 2015, Planning, 6, Task 3

13. d. The information is insufficient to make an assessment.

    The information provided tells us that, as of the fourth month, more money has been spent than was planned. However, we need to know how much work has been completed to determine how the project is performing. In earned value terms, we are missing the EV or the measure of work expressed in terms of the budget authorized for the work. It is the budget associated for the work that has been completed. The objective is to establish progress measurement criteria for each WBS component to measure work in progress. EV then is monitored incrementally to determine current status and cumulatively to determine the long-term performance trends. [Monitoring and Controlling]

    PMI®, *PMBOK® Guide*, 2013, 218
    PMI® *PMP Examination Content Outline,* 2015, Monitoring and Controlling, 9, Task 1

14. b. Use the project schedule as an input to determine budget

    Accurate project performance measurement depends on accurate cost and schedule information. The project schedule includes planned start and finish dates for all activities, milestones, work packages, and control accounts. This information is used to aggregate costs to the calendar period for which the costs are planned to be incurred. The project schedule is an input to the Determine Budget process. [Planning]

    PMI®, *PMBOK® Guide*, 2013, 210
    PMI® *PMP Examination Content Outline,* 2015, Planning, 6, Task 3

15. d. Are usually beyond the project manager's control

    Overhead includes costs such as rent, insurance, or heating, that pertain to the project as a whole and cannot be attributed to a particular work item. The amount of overhead to be added to the project is frequently decided by the performing organization and is beyond the control of the project manager. [Planning]

    Meredith, J.R. and Mantel, Jr., S.J., *Project Management: A Managerial Approach*, 2012, 301
    PMI®, *PMBOK® Guide*, 2013, 202
    PMI® *PMP Examination Content Outline,* 2015, Planning, 6, Task 3

16. d. ETC = EAC – AC

The ETC is the expected cost to finish the remaining work. This formula is used if the work is performing as planned. Otherwise you may need to re-estimate the remaining work from the bottom up. [Monitoring and Controlling]

PMI®, *PMBOK® Guide*, 2013, 224
PMI® *PMP Examination Content Outline,* 2015, Monitoring and Controlling, 9, Task 1

17. c. Order-of-magnitude

An order-of-magnitude estimate, which is referred to also as a ballpark estimate, has an accuracy range of –25% to 75% and is made without detailed data. These estimates are made when the project is in the initiation stage. The cost estimates should be reviewed and refined during the project to reflect additional detail. As the project progresses the accuracy of the estimates will increase. Later in the project, for example the range of accuracy can be narrowed to –5% to +10%. [Planning]

PMI®, *PMBOK® Guide*, 2013, 201
PMI® *PMP Examination Content Outline,* 2015, Planning, 6, Task 3

18. b. Issue a change request

Before a revised cost baseline leading to a budget update can be prepared, it is necessary to issue a change request, which may include preventive or corrective action. These change requests then are reviewed and processed through the Perform Integrated Change Control process. Change requests are an output of the Control Costs process. After analysis of project performance, change requests may be also needed for other components in addition to the cost baseline. [Monitoring and Controlling]

PMI®, *PMBOK® Guide*, 2013, 225
PMI® *PMP Examination Content Outline,* 2015, Monitoring and Controlling, 9, Task 1

19. d. Value engineering

Value engineering considers possible cost trade-offs as a design evolves. The technique entails identifying the functions that are needed and analyzing the cost effectiveness of the alternatives available for providing them. It helps optimize project life cycle costs, save time, increase profits, improve quality, increase market share, solve problems, and contribute toward more effective resource use [Monitoring and Controlling]

PMI®, *PMBOK® Guide*, 2013, 566
PMI® *PMP Examination Content Outline,* 2015, Monitoring and Controlling, 9, Task 1

20. a. CPI

The CPI has been proven to be an accurate and reliable forecasting tool. It is a measure of the cost efficiency of budgeted resources and is considered to be the most critical earned value measurement metric. It measures the cost efficiency for the work completed. Its indices are useful for providing a basis to estimate project cost and schedule outcome. [Monitoring and Controlling]

PMI®, *PMBOK® Guide*, 2013, 219
PMI® *PMP Examination Content Outline,* 2015, Monitoring and Controlling, 9, Task 1

21. b. Performance measurement baseline

The performance measurement baseline is the approved plan for the project work, which is used to compare actual project execution. From it, deviations are measured for management control. It typically integrates scope, schedule, and cost parameters; it may include technical and quality parameters. It includes a contingency reserve but not a management reserve. Contingency reserves are the budget in the cost baseline allocated to identifying risks that are accepted and for which contingent or mitigation responses are developed. [Monitoring and Controlling]

PMI®, *PMBOK® Guide*, 2013, 206, 302, 549
PMI® *PMP Examination Content Outline,* 2015, Monitoring and Controlling, 9, Task 1

22. b. Parametric estimating

Parametric estimating involves using statistical relationships between historical data and other variables to calculate a cost estimate for project work. This approach can produce higher levels of accuracy depending on the sophistication and underlying data in the model. Parametric estimates may be used for the entire project or for parts of it in conjunction with other estimating techniques. The example is representative of a simple parametric model. [Planning]

PMI®, *PMBOK® Guide*, 2013, 205
PMI® *PMP Examination Content Outline,* 2015, Planning, 6, Task 3

23. d. Costs that have been authorized and incurred

These costs are part of work performance data about project progress. In addition, data include information about project progress such as which activities have started, their progress, and which deliverables have finished. Other information includes costs that have been authorized and incurred. Updating the budget requires knowledge about the actual costs spent to date, and any budget changes are approved according to the Perform Integrated Change Control process. [Monitoring and Controlling]

PMI®, *PMBOK® Guide*, 2013, 216–217
PMI® *PMP Examination Content Outline,* 2015, Monitoring and Controlling, 9, Task 1

24. a. Code of accounts allocation provision

Rules of earned value performance measurement are part of the cost management plan and may (1) define the points in the WBS where measurement of control accounts will be performed; (2) establish the EV measurement techniques such as weighted milestones, fixed-formula, percent complete, etc., to be used; and (3) specific tracking methods and EV equations for calculating the EAC forecasts to provide a validity check on the bottom-up EAC. [Planning]

PMI®, *PMBOK® Guide*, 2013, 199
PMI® *PMP Examination Content Outline,* 2015, Planning, 6, Task 3

25. a. EAC = ETC – AC

EAC is the expected total cost of completing all work expressed as the sum of the actual cost to date and the estimate at completion. There are four methods to compute it based on different assumptions. If the CPI is expected to be the same for the remainder of the project use EAC = BAC/CPI or answer b. If future work will be accomplished at the planned rate, use EAC = AC + BAC – EV or answer c. If the initial plan no longer is valid, use EAC = AC + Bottom-up ETC. If both the CPI and the SPI influence the remaining work, use EAC = AC + (BAC-EV)/(CPI x SPI) or answer d. [Monitoring and Controlling]

PMI®, *PMBOK® Guide*, 2013, 220–221
PMI® *PMP Examination Content Outline,* 2015, Monitoring and Controlling, 9, Task 1

26. a. How direct and indirect costs will be handled

The scope statement, as part of the scope baseline, is a key input in the Estimate Costs process and should be reviewed. It provides the product description, acceptance criteria, key deliverables, project boundaries, assumptions, and constraints about the project. It also notes one basic assumption that must be made as costs are estimated is whether the estimates will be limited only to direct project costs or whether they also will include indirect project costs. The indirect costs cannot be directly traced to a single project and are accumulated and allocated equally over multiple projects using an approved and documented accounting procedure. [Planning]

PMI®, *PMBOK® Guide*, 2013, 202
PMI® *PMP Examination Content Outline,* 2015, Planning, 6, Task 3

27. b. PV, EV, and AC

Cumulative cost curves, or S-curves, enable the project manager to monitor the three parameters of planned value, earned value, and actual cost and to report on a period-by-period basis, such as weekly, monthly, or on a cumulative basis. Please see Figure 7-12 in the *PMBOK® Guide*, 2013 for a reference. [Monitoring and Controlling]

PMI®, *PMBOK® Guide*, 2013, 219
PMI® *PMP Examination Content Outline,* 2015, Monitoring and Controlling, 9, Task 1

28. d. Represent the basic level at which project performance is measured and reported

Control accounts represent a management control point where scope, budget, actual costs, and schedule are integrated and compared to earned value for performance measurement. Each control account has a unique code or account number that links directly to the organization's accounting system. They are described in the project's cost management plan. [Planning]

PMI®, *PMBOK® Guide*, 2013, 132, 199, and 533
PMI® *PMP Examination Content Outline,* 2015, Planning, 6, Task 3

29. c. Time reporting systems

Variance analysis, as used in earned value management, is the explanation (cause, issue, and corrective action) for cost and schedule variance at completion (VAC = BAC-EAC). If earned value is not being used, variance analysis compares planned activity cost against actual activity cost to identify variances between the cost baseline and actual project performance. In cost control, variance analysis is important to determine the cause and degree of variance relative to the baseline to decide if corrective or preventive action is required. As more work is done, the percentage of acceptable variances tends to decrease. Trend analysis examines project performance over time to determine performance status to see if it is improving or deteriorating. Earned value performance compares the performance measurement baseline to actual schedule and cost performance. [Monitoring and Controlling]

PMI®, *PMBOK® Guide*, 2013, 222–223
PMI® *PMP Examination Content Outline,* 2015, Monitoring and Controlling, 9, Task 1

30. a. Before the budget is complete

Often project cost estimates are prepared after budgetary approval is provided. However, activity cost estimates should be prepared before the budget is complete. Activity cost estimates are an output of the Estimate Costs process and an Input to the Determine Budget process. They are quantitative assessments of the probable costs required to complete project work. Costs are estimated for all resources applied to the activity cost estimate such as direct labor, materials, equipment, services, facilities, and technology. They also include other categories such as cost of financing (e.g., interest charges), an inflation allowance, exchange rates, or a cost contingency reserve. Indirect costs may be included at the activity level or at a higher level. [Planning]

PMI®, *PMBOK® Guide*, 2013, 210
PMI® *PMP Examination Content Outline,* 2015, Planning, 6, Task 3

31. c. Use a Delphi approach

    In Estimate Costs, one tool and technique is group decision-making techniques, which include brainstorming, the Delphi approach, or nominal group techniques. They are useful to engage stakeholders to improve the accuracy of estimates and commitment to emerging estimates. When a structured group of people, who are close to the technical execution of the work, are involved in estimating the work additional information is obtained, leading to more accurate estimating. When people are involved in the estimating process, they tend to become more committed to meeting the resulting estimates that are made. [Planning]

    PMI®, *PMBOK® Guide*, 2013, 207
    PMI® *PMP Examination Content Outline,* 2015, Planning, 6, Task 3

32. d. Request additional contingency reserves for your budget

    Reserve analysis monitors the status of contingency and management reserves to see if they are still needed or if additional reserves need to be requested. The reserves may be used as planned to cover risk mitigation costs. In this situation, risk identification and analysis are ongoing, and they may indicate a need to request additional reserves to the budget, which may be a management reserve, a contingency reserve, or both. [Monitoring and Controlling]

    PMI®, *PMBOK® Guide*, 2013, 225
    PMI® *PMP Examination Content Outline,* 2015, Monitoring and Controlling, 9, Task 1

33. b. Global supply and demand conditions can influence resource costs

    Market conditions, along with published commercial information, are enterprise environmental assets, an input to Estimate Costs. Market conditions describe the products, services, and results available in the market and who provides them, along with any terms and conditions. As well, regional and global supply and demand conditions can greatly influence resource costs, assisting in the estimating process. [Planning]

    PMI® *PMBOK® Guide*, 2013, 203
    PMI® *PMP Examination Content Outline,* 2015, Planning, 6, Task 3

34. a. CVs are severe, and a realistic measure of performance is needed

After the CVs exceed certain ranges, the original project budget may be questioned and changed as a result of new information. Changes to the cost baseline are incorporated in response to approved changes in scope, activity resources, or cost estimates. [Monitoring and Controlling]

PMI®, *PMBOK® Guide*, 2013, 226
PMI® *PMP Examination Content Outline,* 2015, Monitoring and Controlling, 9, Task 1

35. d. Document lessons learned

Lessons learned but not documented are "lessons lost." The lessons learned knowledge database will help current project members, as well as people on future projects, make better decisions. Accordingly, the reasons for the variance, the rationale supporting the corrective action, and other related information must be documented. They require updates as part of updates to organizational process assets as an output of Control Costs. [Monitoring and Controlling]

PMI®, *PMBOK® Guide*, 2013, 226
PMI® *PMP Examination Content Outline,* 2015, Monitoring and Controlling, 9, Task 6

36. a. Aggregate these estimates by work packages

The WBS provides the relationship among all the project deliverables and their components and should be reviewed before the budget is developed. As the budget is determined, the cost estimates for the activities should be aggregated by the work packages in the WBS. Then, they are aggregated for the control accounts and finally for the entire project. The summation of the control accounts comprises the cost baseline, and it enables a time-phased view of it, which typically is displayed as an S-cure. Refer to Figure 7-9 in the *PMBOK® Guide*, 2013 for an example. [Planning]

PMI®, *PMBOK® Guide*, 2013, 213–214
PMI® *PMP Examination Content Outline,* 2015, Planning, 6, Task 3

37. a. All future work will need to be performed immediately with the range of the TCPI

The TCPI is a measure of cost performance that is required to be achieved with the remaining resources to meet a specified management goal. It is expressed as the ratio of the cost to finish the remaining work to the remaining budget. If the cumulative CPI falls below the baseline, future project work will need to be performed immediately within the range of the TCPI to stay within the authorized budget at completion. Whether this performance level is achievable is a judgment call based on considerations such as risk, schedule, and technical performance. Refer to Figure 7-13 in the *PMBOK® Guide*, 2013, for an example. . [Monitoring and Controlling]

PMI®, *PMBOK® Guide*, 2013, 221
PMI® *PMP Examination Content Outline*, 2015, Monitoring and Controlling, 9, Task 1

38. d. Consider the forecasted EAC

Forecasting is a tool and technique in Control Costs. In this situation, use of the forecasted EAC is recommended because it involves making projections of future conditions and events on the basis of current performance information and other knowledge that is available. Forecasts are generated, updated, and reissued based on work performance data provided as the project is executed. EACs are typically based on the actual costs for work completed plus an estimate to complete. [Monitoring and Controlling]

PMI®, *PMBOK® Guide*, 2013, 220
PMI® *PMP Examination Content Outline*, 2015, Monitoring and Controlling, 9, Task 1

39. b. At –$200, the physical progress is being accomplished at a slower rate than is planned, indicating an unfavorable situation.

SV (schedule variance) is calculated: EV – PV or $1,000 – $1,200 = –$200. Because the SV is negative, physical progress is being accomplished at a slower rate than planned. It is a useful metric because it can indicate when a project is falling behind or is ahead of its baseline schedule. Ultimately, it equals zero when the project is complete because all the planned values have been earned. It should be used along with monitoring the critical path. [Monitoring, and Controlling]

PMI®, *PMBOK® Guide*, 2013, 218, 224
PMI® *PMP Examination Content Outline*, 2015, Monitoring and Controlling, 9, Task 1

40. d. Place emphasis on improving the productivity by which work was being performed

CPI = EV/AC and measures the efficiency of the physical progress accomplished compared to the baseline. A CPI of 0.84 means that for every dollar spent, you are only receiving 84 cents of progress. Therefore, you should focus on improving the productivity by which work is being performed as now it represents a cost overrun for the project. [Monitoring and Controlling]

PMI®, *PMBOK® Guide*, 2013, 219, 224
PMI® *PMP Examination Content Outline,* 2015, Monitoring and Controlling, 9, Task 1

Project Quality Management

Study Hints

The Project Quality Management questions on the PMP® certification exam are straightforward—especially if you know definitions of terms and understand statistical process control. You are not required to solve quantitative problems, but there may be questions on statistical methods of measuring and controlling quality.

The exam is likely to reflect a heavy emphasis on customer satisfaction and continuous improvement through the use of quality tools such as Pareto analysis and cause-and-effect diagrams. You should know the differences among Plan Quality Management, Perform Quality Assurance, and Perform Quality Control.

The *PMBOK® Guide* includes all quality-related activities under the term Project Quality Management, which comprises the three quality processes mentioned above. Before taking the practice test, review *PMBOK® Guide, 2013,* Figure 8-1 for an overview of the Project Quality Management structure. Know this chart thoroughly.

Following is a list of the major Project Quality Management topics. Use it to help focus your study efforts on the areas most likely to appear on the exam.

Major Topics

Key *PMBOK® Guide* concepts
Quality defined
Quality management
Quality policy
Quality and grade
Accuracy and precision
Customer satisfaction
Prevention over inspection
Continuous improvement
SIPOC model
Management responsibility
Seven basic quality tools

- Cause-and-effect diagrams
- Flowcharts
- Checksheets
- Pareto diagrams
- Histograms
- Control charts
- Scatter diagrams

Plan quality management tools

- Cost-benefit analysis
- Benchmarking
- Design of experiments
- Cost of quality
- Statistical sampling
- Brainstorming
- Force field analysis
- Nominal group technique

Key quality planning documents

- Quality management plan
- Quality metrics
- Quality checklists
- Process improvement plan

Quality control measurements
Quality management and control tools

- Affinity diagrams
- Process decision program charts
- Interrelationship diagraphs
- Tree diagrams
- Prioritization matrices
- Activity network diagrams
- Matrix diagrams

Quality audits
Process analysis
Quality control

- Variable sampling
- Attribute sampling
- Tolerances and control limits
- Prevention
- Probability
- Standard deviation
- Validated changes
- Verified deliverables

Approved change requests review
Impact of motivation on quality
Priority of quality versus cost and schedule
Design and quality

Practice Questions

INSTRUCTIONS: Note the most suitable answer for each multiple-choice question in the appropriate space on the answer sheet.

1. Quality is very important to your company. Each project has a quality statement that is consistent with the organization's vision and mission. Both internal and external quality assurance are provided on all projects to—
   a. Ensure confidence that the project will satisfy relevant quality standards
   b. Monitor specific project results to note whether they comply with relevant quality standards
   c. Identify ways to eliminate causes of unsatisfactory results
   d. Use inspection to keep errors out of the process

2. Benchmarking is a technique used in—
   a. Inspections
   b. Root cause analysis
   c. Plan Quality Management
   d. Control Quality

3. In quality management, the practice "rework" is—
   a. Acceptable under certain circumstances
   b. An adjustment made that is based on quality control measurements
   c. Action taken to bring a defective or nonconforming component into compliance
   d. Not a concern if errors are detected early

4. Requirements documentation is useful in project quality management because it—
   a. Provides better product definition and product development
   b. Helps products to succeed in the marketplace
   c. Helps plan how quality control will be implemented on the project
   d. Helps control the overall cost of quality

5. Often, one quality tool may lead to the development of another useful quality tool. An example is—
   a. Process decision point program charts
   b. A roadmap
   c. Interrelationship diagraphs
   d. Force field analysis

6. You are leading a research project that will require between 10 and 20 aerospace engineers. Some senior-level aerospace engineers are available. They are more productive than junior-level engineers, who cost less and who are available as well. You want to determine the optimal combination of senior- and junior-level personnel. In this situation, the appropriate technique to use is to—

   a. Conduct a design of experiments
   b. Use the Ishikawa diagram to pinpoint the problem
   c. Prepare a control chart
   d. Analyze the process using a Pareto diagram

7. Checksheets are one of the seven basic quality tools. Often overlooked is that the data about frequencies or consequences in checksheets—

   a. Can be used to formulate the problem for a cause-and-effect diagram
   b. Use statistical techniques to compute a "loss function" to determine the cost of producing products that fail to achieve a target value
   c. Can be displayed in Pareto diagrams
   d. Are useful for benchmarking

8. One area that often influences perceptions of quality is—

   a. Cultural concerns
   b. Information from the risk register
   c. Organizational quality policies
   d. Specific tolerances

9. Which of the following statements best describes attribute sampling versus variables sampling?

   a. Attribute sampling is concerned with prevention, whereas variables sampling is concerned with inspection.
   b. Attribute sampling is concerned with conformance, whereas variables sampling is concerned with the degree of conformity.
   c. Attribute sampling is concerned with special causes, whereas variables sampling is concerned with any causes.
   d. Both are the same concept.

10. Your project scheduler has just started working with your project and has produced defect reports for the past two accounting cycles. If this continues, these defect reports could provide the potential for customer dissatisfaction and lost productivity that is due to rework. You discovered that the project scheduler needs additional training on using the scheduling tool that is used on your project. The cost of training falls under which one of the following categories?

    a. Overhead costs
    b. Failure costs
    c. Prevention costs
    d. Indirect costs

11. In order to use approved change requests as an input to Control Quality, it is important to—

    a. Have a change management plan
    b. Update the change log
    c. Implement recommendations from the CCB
    d. Verify their timely implementation

12. It is important to use statistical sampling during Plan Quality Management in order to—

    a. Help determine the cost of quality
    b. Determine when attribute sampling versus variables sampling is more appropriate
    c. Set tolerances and control limits
    d. Differentiate between special causes and random causes

13. Rank ordering of defects should be used to guide corrective action. This is the underlying principle behind—

    a. Trend analysis
    b. Inspections
    c. Control charts
    d. Pareto diagrams

14. Project quality management was once thought to include only inspection or quality control. In recent years, the concept of project quality management has broadened. Which statement is NOT representative of the new definition of quality management?

    a. Quality is designed into the product or service, not inspected into it.
    b. Quality is the concern of the quality assurance staff.
    c. Customers require a documented and, in some cases, registered quality assurance system.
    d. National and international standards and guidelines for quality assurance systems are available.

15. Assume your organization is a start-up company, and you are trying to explain the importance of project quality management to the management team, which has not worked in this area before. You make an analogy to archery in your presentation, because one of the managers is an expert in it. You tell them arrows clustered tightly in one area of the target, even if they are not in the area of the bull's eye, have a high precision, but targets where the arrows are more spread out but are equidistant from the bull's eye are considered to have the same degree of accuracy. This analogy shows—

    a. Precision measures exactness, while accuracy assesses correctness
    b. Both are of equal importance in preparing and executing a quality management plan on the company's projects
    c. The next step is to design an experiment to determine the sufficient degree of accuracy and precision required for an upcoming project
    d. Precise measurements are accurate measurements, but accurate measurements may not necessarily be precise ones

16. Your quality assurance department recently performed a quality audit of your project and identified a number of findings and recommendations. One recommendation seems critical and should be implemented because it affects successful delivery of the product to your customer. Your next step should be to—

    a. Call a meeting of your project team to see who is responsible for the problem
    b. Reassign the team member who had responsibility for oversight of the problem
    c. Perform product rework immediately
    d. Issue a change request to implement the needed corrective action

17. Using control charts is one way to monitor cost and schedule variances, volume, frequency of scope changes, and other results to determine if the project management processes are stable. Assume you are working on a project that implements a repetitive process because its result is used many times after the process is designed and tested in telecommunications. Since it is repetitive, the control limits are:

    a. Seven consecutive plot points above the mean
    b. Ways to identify which factors have the greatest influence
    c. +/− 3 s around a process mean set at 0 s
    d. Seven consecutive plot points below the mean

18. Projects tend to result in change, and people tend to resist change. Wanting to avoid resistance against the new manufacturing process that you are implementing, you—

    a. Prepare and follow a change management plan
    b. Use force field analysis
    c. Involve all stakeholders using a nominal group technique approach
    d. Use benchmarking

19. Assume previous projects in the organization have overrun their budgets consistently and tend to require more contingency reserves than in the original budget. You are striving to avoid the need for additional contingency on your project and are doing so by—

    a. Involving more stakeholders in the risk identification process
    b. Using a process decision program chart
    c. Consulting the knowledge transfer system
    d. Using quality circles

20. Quality inspections also may be called—

    a. Control tests
    b. Walkthroughs
    c. Statistical sampling
    d. Checklists

21. Your management has prescribed that a quality audit be conducted at the end of every phase in a project. This audit is part of the organization's—

    a. Perform Quality Assurance process
    b. Control Quality process
    c. Quality improvement program
    d. Process adjustment program

22. You are managing a major international project, and your contract requires you to prepare both a project plan and a quality management plan. Your core team is preparing a project quality management plan. Your first step in developing this plan is to—

 a. Determine specific metrics to use in the quality management process
 b. Identify the quality standards for the project
 c. Develop a quality policy for the project
 d. Identify specific quality management roles and responsibilities for the project

23. Recently your company introduced a new set of "metal woods" to its established line of golfing equipment. As you work in quality assurance on this project, you decide to use a matrix diagram as it—

 a. Analyzes the product development cycle after product release to determine strengths and weaknesses
 b. Shows the strength of relationships between factors, causes, and objectives
 c. Identifies non-conformity, gaps, and shortcomings
 d. Analyzes the quality of the processes of the project against organizational standards

24. An often overlooked function of a quality audit is to—

 a. Confirm the implementation of approved change requests
 b. Examine constraints experienced
 c. Update training plans
 d. Establish quality metrics

25. On-time performance, cost control, defect frequency, failure rates, availability, reliability, and test coverage are examples of—

 a. Incentives to vendors to make quality commitments to improve overall performance
 b. Quality metrics as an input to Perform Quality Assurance
 c. Methods that usually result in lower costs and increased profitability
 d. Items to include as goals in the quality management plan

26. Rework required, causes for rejection, and the need for process adjustment are examples of—

 a. Work performance data in Perform Quality Assurance
 b. Work performance information in Control Quality
 c. Items that are part of the process improvement plan
 d. Items to cover at meetings used to prepare the quality management and process improvement plans

27. Quality control measurements are captured—
    a. As specified in the Plan Quality Management process
    b. To prepare an operational definition
    c. To prepare a control chart
    d. As an output to Perform Quality Assurance

28. The control chart is a tool used primarily to help—
    a. Monitor process variation over time
    b. Measure the degree of conformance
    c. Determine whether results conform to benefits
    d. Determine whether results conform to requirements

29. An often, but important, overlooked output of the Plan Quality Management process is the Process Improvement Plan. While its focus is on the steps to analyze project management and product development processes to identify activities to enhance their value, it also includes—
    a. An approach to ensure the information for decisions is of the highest reliability and accuracy
    b. A process to ensure checklists do not limit the Plan Quality Assurance and Control Quality process
    c. A process configuration with interfaces
    d. A process to focus on the project's value proposition

30. You are a project manager for residential construction. As a project manager, you must be especially concerned with building codes—particularly in the Plan Quality Management process. You must ensure that building codes are reflected in your project plans because—
    a. Standards and regulations are an input to Plan Quality Management
    b. Quality audits serve to ensure there is compliance with regulations
    c. They are a cost associated with quality initiatives
    d. Compliance with standards is the primary objective of Control Quality

31. You work as a project manager in the largest hospital in the region. Studies have shown that patients have to wait for long periods before being treated. To assist in identifying the factors contributing to this problem, you and your team have decided to use which of the following techniques?
    a. Cause-and-effect diagrams
    b. Pareto analysis
    c. Scatter diagrams
    d. Control charts

32. Basic quality management as defined in the *PMBOK® Guide* is intended to be compatible with the International Organization for Standardization (ISO) quality standards. This means that, among other things, every project should—

    a. Describe how products should be produced
    b. State specifics for the implementation of quality systems should be stated
    c. Have a framework for quality systems
    d. Have a quality management plan

33. All but one of the following is objectives of a quality audit—

    a. Good practices being implemented are identified
    b. Gaps and shortcomings are identified
    c. Good practices can be introduced elsewhere
    d. Root cause analysis has been performed as part of process analysis

34. Quality checklists are an output of Plan Quality Management. These checklists—

    a. Form the basis for quality control measurements
    b. Are a tool and technique used in the quality audits
    c. Incorporate the acceptance criteria included in the scope baseline
    d. Contain product and project attributes

35. One way to display the sequence of steps and the branching possibilities that exist for a process that transforms one or more inputs into one or more outputs is to use a process map or a—

    a. Checklist
    b. Flowchart
    c. Tree diagram
    d. Process decision program chart

36. Quality requirements of the project are recorded in—

    a. Process improvement plan
    b. Quality management plan
    c. Quality baseline
    d. Quality metrics

37. The below Pareto chart indicates defects in areas associated with billing a client for project services. Based on this Pareto analysis, which area, or areas, indicate the greatest opportunity for improvement?

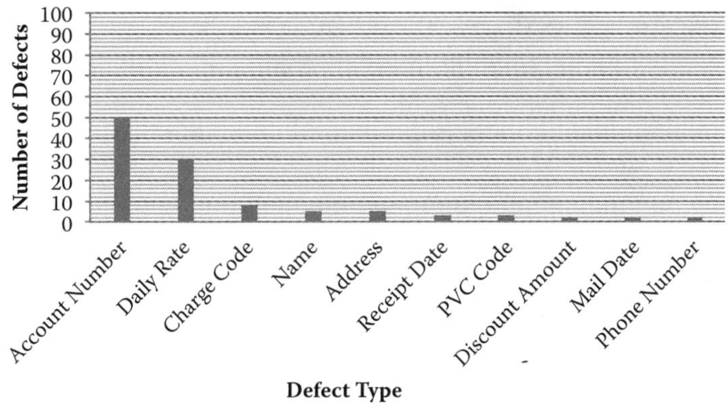

a. The account number, because if it is incorrect, the invoice may be sent to the wrong client.
b. The daily rate, because if it is incorrect, the total amount of the invoice will be wrong, which impacts the cash flow.
c. The charge code, name, address, receipt date, pvc code, discount amount, mail date, and phone number, because they are fairly easy to confirm and correct, thereby significantly reducing the types of defects.
d. The account number and daily rate, because they account for 80 percent of all defects.

38. You have decided to use a fishbone diagram to identify the relationship between an effect and its causes. To begin, you should first—

a. Select an interdisciplinary team who has used the technique before to help brainstorm the problem
b. Determine the major categories of defects
c. Set up a process analysis using HIPO charts
d. Identify the problem

39. Correlation charts seek to explain a change in the dependent variable, Y, in relationship to a change in the corresponding independent variable, X. They are also known as a(n)—

a. Histogram
b. Attribute chart
c. Control chart
d. Scatter diagram

40. The quality management plan should be reviewed early because—

    a. It will show any boundaries to consider
    b. It will enable a sharper focus on the project's value proposition
    c. It can determine whether additional stakeholders, who were not involved in the planning meetings, should be consulted
    d. It will focus on the process metrics for greater efficiency

Answer Sheet

1.	a	b	c	d	21.	a	b	c	d
2.	a	b	c	d	22.	a	b	c	d
3.	a	b	c	d	23.	a	b	c	d
4.	a	b	c	d	24.	a	b	c	d
5.	a	b	c	d	25.	a	b	c	d
6.	a	b	c	d	26.	a	b	c	d
7.	a	b	c	d	27.	a	b	c	d
8.	a	b	c	d	28.	a	b	c	d
9.	a	b	c	d	29.	a	b	c	d
10.	a	b	c	d	30.	a	b	c	d
11.	a	b	c	d	31.	a	b	c	d
12.	a	b	c	d	32.	a	b	c	d
13.	a	b	c	d	33.	a	b	c	d
14.	a	b	c	d	34.	a	b	c	d
15.	a	b	c	d	35.	a	b	c	d
16.	a	b	c	d	36.	a	b	c	d
17.	a	b	c	d	37.	a	b	c	d
18.	a	b	c	d	38.	a	b	c	d
19.	a	b	c	d	39.	a	b	c	d
20.	a	b	c	d	40.	a	b	c	d

Answer Key

1. a. Ensure confidence that the project will satisfy relevant quality standards

   Quality assurance increases project effectiveness and efficiency and provides added benefits to project stakeholders. It implements the planned and systematic acts and processes in the project's quality management plan. It builds confidence that an unfinished output or a work in progress will be completed in a way to meet specified requirements and specifications. Quality assurance should be performed throughout the project. [Executing]

   PMI®, *PMBOK® Guide*, 2013, 227, 242–243
   PMI® *PMP Examination Content Outline,* 2015, Executing, 8, Task 3

2. c. Plan Quality Management

   Benchmarking involves comparing actual or planned practices to those practices of comparable projects to identify best practices, to note ideas for improvement, and to provide a way to measure performance. Benchmarking may be done by the performing organization, external to it, or within the same application area. It enables analogies from projects in different application areas to be made. It is a tool and technique used in Plan Quality Management. [Planning]

   PMI®, *PMBOK® Guide*, 2013, 239
   PMI® *PMP Examination Content Outline,* 2015, Planning, 6, Task 8

3. c. Action taken to bring a defective or nonconforming component into compliance

   Rework is a frequent cause of project overruns. The project team must make every reasonable effort to control and minimize rework so that defective or nonconforming components are brought into compliance with requirements or specifications. Less rework is a primary benefit of meeting quality requirements. It also is an example of an internal failure cost or a cost of nonconformance and represents money spent during and after the project because of failure. [Monitoring and Controlling]

   PMI®, *PMBOK® Guide*, 2013, 235, 559
   PMI® *PMP Examination Content Outline,* 2015, Monitoring and Controlling, 9, Task 1

4. c. Helps plan how quality control will be implemented on the project

Requirements documentation is an input to Plan Quality Management. It captures the requirements the project should meet regarding stakeholder expectations. Requirements documentation components include project and product quality requirements, and the project team uses the requirements to help plan how the project will implement quality control. [Planning]

PMI®, *PMBOK® Guide*, 2013, 234
PMI® *PMP Examination Content Outline*, 2015, Planning, 6, Task 8

5. c. Interrelationship diagraphs

They are an adaptation of relationship diagraphs and provide a process for creative problem solving in moderately complex scenarios that have intertwined logical relationships for up to 50 relevant items. They are often developed from data in other tools such as the affinity diagram, the tree diagram, or the fishbone diagram. [Executing]

PMI®, *PMBOK® Guide*, 2013, 245
PMI® *PMP Examination Content Outline*, 2015, Executing, 8, Task 3

6. a. Conduct a design of experiments

This technique is used to identify which variables have the most influence on a product or process under development or in production. It also can be used to determine the number and type of tests and their impact on the cost of quality. They also play a role in optimizing products or processes. It is a tool and technique in Plan Quality Management. For example, roller blade designers might want to determine which combination of number of wheels and titanium ball bearings would produce the most desirable "ride" characteristics at a reasonable cost. This technique, however, can be applied to project management issues such as cost and schedule trade-offs. An appropriately designed "experiment" often will help project managers to find an optimal solution from a relatively limited number of options, and often it help to determine the number and type of tests to use and their impact on quality. It provides a statistical framework to systematically change all of the important factors rather than focusing on factors one at a time. By analyzing the data it can help provide the optimal conditions for the product or process, highlight the factors that influence the results, and reveal the presence of interactions and synergy among the factors. [Planning]

PMI®, *PMBOK® Guide*, 2013, 239–240
PMI® *PMP Examination Content Outline*, 2015, Planning, 6, Task 8

7. c. Can be displayed in Pareto diagrams

   As one of the seven basic quality tools, checksheets, which may be called tally sheets, may be used as a checklist in gathering data. They serve to organize facts in a manner to facilitate the effective collection of useful data about a possible quality problem. They also are used to gather attribute data while performing inspections to identify defects. The data about the frequencies or consequences of defects collected in them are displayed often in Pareto diagrams. [Planning]

   PMI®, *PMBOK® Guide*, 2013, 237
   PMI® *PMP Examination Content Outline,* 2015, Planning, 6, Task 8

8. c. Cultural concerns

   Enterprise environmental factors are an input to Plan Quality Management, one of which is cultural perceptions that may influence expectations about quality. Other enterprise environmental factors are: government agency regulations; rules, standards, and guidelines specific to the application area of the project; and working or operating conditions of the project or its deliverables that may affect product quality. [Planning]

   PMI®, *PMBOK® Guide*, 2013, 234
   PMI® *PMP Examination Content Outline,* 2015, Planning, 6, Task 8

9. b. Attribute sampling is concerned with conformance, whereas variables sampling is concerned with the degree of conformity.

   Attribute sampling determines whether a result does or does not conform. Variables sampling rates a result on a continuous scale to measure the degree of conformity. [Monitoring and Controlling]

   PMI®, *PMBOK® Guide*, 2013, 250
   PMI® *PMP Examination Content Outline,* 2015, Monitoring and Controlling, 9, Task 1

10. c. Prevention costs

Prevention costs include any expenditure directed toward ensuring that quality is achieved the first time. The objective is to focus on cost of conformance and spend money during the project to avoid failures. The cost of quality is a tool and technique in Plan Quality Management and includes all costs over the life of the product in preventing nonconformance to requirements; appraising the product, service, or result for conformance to requirements; and failing to meet requirements leading to rework. [Planning]

PMI®, *PMBOK® Guide*, 2013, 235
PMI® *PMP Examination Content Outline,* 2015, Planning, 6, Task 8

11. d. Verify their timely implementation

Approved change requests are an input to Control Quality. As part of the Perform Integrated Change Control process, a change-log update shows some changes are approved, while others are rejected or deferred. The approved change requests may include modifications such as defect repairs, revised work methods, and a revised schedule. Their timely implementation should be verified. [Monitoring and Controlling]

PMI®, *PMBOK® Guide*, 2013, 251
PMI® *PMP Examination Content Outline,* 2015, Monitoring and Controlling, 9, Task 1

12. a. Help determine the cost of quality

Statistical sampling is a tool and technique in Plan Quality Management in that it involves choosing part of a population for inspection such as selecting ten engineering drawings from a population of 95. Sample frequency and sizes are determined at this time so the cost of quality will include items such as the number of tests and expected scrap. Statistical sampling is also a tool and technique in Control Quality. [Planning and Monitoring and Controlling]

PMI®, *PMBOK® Guide*, 2013, 240, 252
PMI® *PMP Examination Content Outline,* 2015, Monitoring and Controlling, 9, Task 1
PMI® *PMP Examination Content Outline,* 2015, Planning, 6, Task 8

13. d. Pareto diagrams

Pareto diagrams are histograms, ordered by frequency of occurrence, that show how many results were generated by type or category of identified cause. The project team should take action to fix the problems that are causing the greatest number of defects first. Pareto diagrams are based on Pareto's Law, which holds that a relatively small number of causes will typically produce a large majority of defects, also called the "solzo rule." The categories on the horizontal axis exist as a valid probability distribution that accounts for about 100% of the possible observations. The relative frequency of each specified cause listed on the horizontal axis decrease in magnitude until the defect source, "other", accounts for any non-specified causes. The Pareto diagram is organized into categories that measure frequencies or consequences. They are one of the seven basic quality tools used in Plan Quality Management. They are also a tool and technique in Control Quality. [Planning and Monitoring and Controlling]

PMI®, *PMBOK® Guide*, 2013, 237, 252
PMI® *PMP Examination Content Outline,* 2015, Monitoring and Controlling, 9, Task 1
PMI® *PMP Examination Content Outline,* 2015, Planning, 6, Task 8

14. b. Quality is the concern of the quality assurance staff.

Quality concerns all levels of management and staff. Its success requires participation from all members of the project team with management providing the needed resources to succeed. It recognizes the importance of customer satisfaction, prevention over inspection, continuous improvement, and the cost of quality. The project management team determines the level of precision and accuracy to include in the project quality management plan. [Planning, Executing, and Monitoring and Controlling]

PMI®, *PMBOK® Guide*, 2013, 227–229
PMI® *PMP Examination Content Outline,* 2015, Planning, 6, Task 8
PMI® *PMP Examination Content Outline,* 2015, Executing, 8, Task 3
PMI® *PMP Examination Content Outline,* 2015, Monitoring and Controlling, 9, Task 1

15. a. Precision measures exactness, while accuracy assesses correctness

In project quality management, the project team determines the level of precision and the level of accuracy needed for the project and documents it in the project's quality management plan. Precise measurements are not necessarily accurate ones, and accurate measurements are not necessarily precise ones. [Planning]

PMI®, *PMBOK® Guide*, 2013, 228
PMI® *PMP Examination Content Outline,* 2015, Planning, 6, Task 8

16. d. Issue a change request to implement the needed corrective action

The information obtained from a quality audit can be used to improve quality systems and performance. The subsequent effort to correct any deficiencies should result in a reduced cost of quality and an increase in customer satisfaction. In most cases, implementing quality improvements requires preparation of change requests, which are an output of the Perform Quality Assurance process. They are used to take corrective action preventive action or perform defect repair and are an input to the Perform Integrated Change Control process. [Executing]

PMI®, *PMBOK® Guide*, 2013, 247
PMI® *PMP Examination Content Outline,* 2015, Executing, 8, Task 3

17. c. +/– 3 s around a process mean set at 0 s

Control charts are one of the seven basic quality tools used as an input in Plan Quality Management and Control Quality. The key words in this question are "repetitive process", which means the control limits are set at +/– 3 s around a process mean set at 0 s. The control charts are used to determine whether or not a process is stable or has predictable performance. Upper and lower limits are set based on requirements, and they reflect the maximum or minimum values allowed. These control limits are not specification limits. They are determined statistically. The project manager and others can use the statistically calculated control limits to identify the points where corrective action is required. [Planning and Monitoring and Controlling]

PMI®, *PMBOK® Guide*, 2013, 238, 252
PMI® *PMP Examination Content Outline,* 2015, Planning, 6, Task 8
PMI® *PMP Examination Content Outline,* 2015, Monitoring and Controlling, 9, Task 1

18. b. Use force field analysis

    Force field analysis is a tool and technique used in Plan Quality Management. Its purpose is to use diagrams of the forces for and against change to better plan for change from the beginning. It then can be used by the project manager and the team as a decision-making technique. Developed by Kurt Lewin, it shows the positive or driving forces for change and the negative or obstacle, restraining forces. The project team then can focus effort in quality planning on ways to strengthen the driving, positive forces or weaken the negative ones. [Planning]

    PMI®, *PMBOK® Guide*, 2013, 240
    PMI® *PMP Examination Content Outline,* 2015, Planning, 6, Task 8

19. b. Using a process decision program chart

    A process decision program chart is a tool and technique in Plan Quality Management and in Perform Quality Assurance. It is used to understand a goal, in this example the need to avoid budget overruns and any additional contingency reserve, in relation to the steps to achieve the goal. These charts are a method for contingency planning because they aid the team in anticipating the intermediate steps that could derail achievement of the goal. [Planning and Executing]

    PMI®, *PMBOK® Guide*, 2013, 245
    PMI® *PMP Examination Content Outline,* 2015, Planning, 6, Task 8
    PMI® *PMP Examination Content Outline,* 2015, Executing, 8, Task 3

20. b. Walkthroughs

    Inspections comprise an examination of a work product to determine if it conforms to standards. Additional names for inspections are walkthroughs, audits, reviews, or peer reviews (in some application areas, these terms may have narrow and specific meanings). They are a tool and technique in Control Quality. [Monitoring and Controlling]

    PMI®, *PMBOK® Guide*, 2013, 252
    PMI® *PMP Examination Content Outline,* 2015, Monitoring and Controlling, 9, Task 1

21. a. Perform Quality Assurance process

    Quality assurance involves auditing the quality requirements from quality control measurements to see if appropriate quality standards and operational definitions are used. It facilitates improvements in the quality processes of a project. Quality audits are tools and techniques in the Perform Quality Assurance process and are structured, independent processes to determine if project activities comply with organizational and project policies, processes, and procedures. [Executing]

    PMI®, *PMBOK® Guide*, 2013, 242, 247
    PMI® *PMP Examination Content Outline,* 2015, Executing, 8, Task 3

22. c. Develop a quality policy for the project

    The quality policy includes the overall intentions and direction of the organization with regard to quality, as formally expressed by top management. If the performing organization lacks a formal quality policy or if the project involves multiple performing organizations, as in a joint venture or a consortium, the project management team must develop a quality policy for the project. The quality policy is one of the organizational process assets used as an input in Plan Quality Management, which is endorsed by executives. An output of this process is the quality management plan that describes how the organization's quality policies will be implemented. [Planning]

    PMI®, *PMBOK® Guide*, 2013, 234, 241
    PMI® *PMP Examination Content Outline,* 2015, Planning, 6, Task 8

23. c. Identifies non-conformity, gaps, and shortcomings

    Matrix diagrams are an example of a quality management and control tool used in the Perform Quality Assurance process. They perform data analysis within the organizational structure identified in the matrix. The matrix diagram then shows the relationships between factors, causes, and objectives that exist between the rows and columns that form the matrix. [Executing]

    PMI®, *PMBOK® Guide*, 2013, 246
    PMI® *PMP Examination Content Outline,* 2015, Executing, 8, Task 3

24. a. Confirm the implementation of approved change requests

    The quality audit is a tool and technique in Perform Quality Assurance. It has many objectives, and the subsequent effort to correct any deficiencies noted should reduce the cost of quality and increase sponsor and customer satisfaction. However, these audits can confirm the implementation of approved change requests including updates, corrective actions, defect repairs, and preventive actions. [Executing]

    PMI®, *PMBOK® Guide*, 2013, 247
    PMI® *PMP Examination Content Outline,* 2015, Executing, 8, Task 3

25. b. Quality metrics as an input to Perform Quality Assurance

    These examples of metrics describe a project or product attribute and how the Control Quality process will measure it. The measurement is an actual value. The tolerance defines the allowable variation to the metric. For example, if an objective is to remain within +/– of the approved budget, the quality metric is used to measure the cost of each deliverable and its percent variation from the approved budget for the deliverable. They are an output of the Plan Quality Management process. [Planning and Executing]

    PMI®, *PMBOK® Guide*, 2013, 242, 252
    PMI® *PMP Examination Content Outline,* 2015, Executing, 8, Task 3
    PMI® *PMP Examination Content Outline,* 2015, Planning, 6, Task 8

26. b. Work performance information in Control Quality

    In the Control Quality process, work performance information is an output. It is the performance data collected from various controlling processes analyzed in context and integrated across areas. Examples include information about project requirements fulfillment as shown in the answer to the question. [Monitoring and Controlling]

    PMI®, *PMBOK® Guide*, 2013, 253
    PMI® *PMP Examination Content Outline,* 2015, Monitoring and Controlling, 9, Task 1

27. a. As specified in the Plan Quality Management process

    One Control Quality output results in measurements that are used as inputs to the Perform Quality Assurance process. These quality control measurements are the documented results of the Control Quality activities. They should be captured in a format specified in the quality management plan. [Monitoring and Controlling]

    PMI®, *PMBOK® Guide*, 2013, 249, 252
    PMI® *PMP Examination Content Outline,* 2015, Monitoring and Controlling, 9, Task 1

28. a. Monitor process variation over time

Used to monitor process variation (i.e., stable or has predictable performance) and to detect and correct changes in process performance, the control chart helps people understand and control their processes and work. It enables the project manager, along with appropriate stakeholders, to identify points where corrective action can be taken to prevent unnatural performance as there may be penalties associated with exceeding the specified limits. They are one of the seven basic quality tools and are an input to the Plan Quality Management process and a tool and technique in the Control Quality process. [Planning and Monitoring and Controlling]

PMI®, *PMBOK® Guide*, 2013, 238, 252
PMI® *PMP Examination Content Outline,* 2015, Planning, 6, Task 8
PMI® *PMP Examination Content Outline,* 2015, Monitoring and Controlling, 9, Task 1

29. c. A process configuration with interfaces

The process configuration is used to facilitate analysis. Other items in this plan are process boundaries, process metrics, and targets for improved performance. Process analysis, a tool and technique in Perform Quality Assurance, follows the steps in this plan to identify needed improvements. [Planning and Executing]

PMI®, *PMBOK® Guide*, 2013, 241, 247
PMI® *PMP Examination Content Outline,* 2015, Executing, 8, Task 3
PMI® *PMP Examination Content Outline,* 2015, Planning, 6, Task 8

30. a. Standards and regulations are an input to Plan Quality Management

During the Plan Quality Management process, the project management team should consider enterprise environmental factors especially when relevant to a specific application area. Other examples are rules and standards, working or operating conditions, and cultural perceptions that may influence quality expectations. These are conditions outside of the control of the project manager and project team. [Planning]

PMI®, *PMBOK® Guide*, 2013, 28, 234
PMI® *PMP Examination Content Outline,* 2015, Planning, 6, Task 8

31. a. Cause-and-effect diagrams

Cause-and-effect diagrams, also called Ishikawa diagrams or fishbone diagrams, are used to illustrate how various causes and sub causes interact to create a special effect. It is named for its developer, Kaoru Ishikawa. These diagrams are useful in linking the undesirable effects seen as special variation to the assigned cause, enabling project teams to implement corrective actions to eliminate the special variation shown in a control chart. They are one of the seven basic quality tools, a tool and technique in Plan Quality Management and in Control Quality. [Planning and Monitoring and Controlling]

PMI®, *PMBOK® Guide*, 2013, 236, 252
PMI® *PMP Examination Content Outline,* 2015, Planning, 6, Task 8
PMI® *PMP Examination Content Outline,* 2015, Monitoring and Controlling, 9, Task 1

32. d. Have a quality management plan

ISO quality standards state every project should have a quality management plan and should have data to show conformance with the plan. Additionally to achieve compatibility with ISO, quality management approaches seek to minimize variation and deliver results that meet detailed requirements. The approach thus recognizes the importance of customer satisfaction, prevention over inspection, continuous improvement, management responsibility, and the cost of quality. [Planning]

PMI®, *PMBOK® Guide*, 2013, 228–229
PMI® *PMP Examination Content Outline,* 2015, Planning, 6, Task 8

33. d. Root cause analysis has been performed as part of process analysis

The purpose of the quality audit is to determine if project activities comply with organizational policies, procedures, and processes. The quality audit has a number of objectives associated with it, but root cause analysis is part of process analysis to identify needed improvements in the process improvement plan. The quality audit objectives include: identifying good and best practices that are being implemented; identifying all nonconformities, gaps, and shortcomings; sharing good practices to similar projects in the organization and possibly in the industry; proactively offering assistance to improve implementation of processes to help the team increase productivity; and highlighting the contributions of each audit in the knowledge repository. [Executing]

PMI®, *PMBOK® Guide*, 2013, 247
PMI® *PMP Examination Content Outline,* 2015, Executing, 8, Task 3

34. c. Incorporate the acceptance criteria included in the scope baseline

As an output of Plan Quality Management, quality checklists are used to verify that a set of required steps has been performed. They are based on the project's requirements and practices and may be simple or complex. They should incorporate the acceptance criteria included in the scope baseline. [Planning]

PMI®, *PMBOK® Guide*, 2013, 242
PMI® *PMP Examination Content Outline,* 2015, Planning, 6, Task 8

35. b. Flowchart

A flowchart is one of the seven basic quality tools, a tool and technique in Plan Quality Management and Control Quality. They are also called process maps as noted in the question. They show the activities, decision points, branching loops, parallel paths, and the overall order of processing by mapping the organizational details of procedures that exist within a horizontal chain of a supplier, input, process, output, and customer (SIPOC) model. [Planning and Monitoring and Controlling]

PMI®, *PMBOK® Guide*, 2013, 236, 252
PMI® *PMP Examination Content Outline,* 2015, Monitoring and Controlling, 9, Task 1
PMI® *PMP Examination Content Outline,* 2015, Planning, 6, Task 8

36. b. Quality management plan

The quality management plan describes how the team will implement the quality policy and how the team will meet the quality requirements set for the project. The style and details of the quality management plan are determined by the quality requirements of the project. [Planning]

PMI®, *PMBOK® Guide*, 2013, 241, 557
PMI® *PMP Examination Content Outline,* 2015, Planning, 6, Task 8

37. d. The account number and daily rate, because they account for 80 percent of all defects.

    Pareto analysis focuses on what Joseph Juran called the vital few. It is named after Vilfredo Pareto, an Italian economist, whose studies showed that 80 percent of the wealth was held by 20 percent of the population. The Pareto diagram is used to identify the vital few sources that are responsible for causing most of the problems effects, in this case the items in the answer. It typically shows that 80 percent of all the problems (defects) are found in 20 percent of the items or areas studied. It is one of the seven basic quality tools, a tool and technique used in Plan Quality Management and Control Quality. [Planning and Monitoring and Controlling]

    PMI®, *PMBOK® Guide*, 2013, 237, 252, 548
    PMI® *PMP Examination Content Outline,* 2015, Monitoring and Controlling, 9, Task 1
    PMI® *PMP Examination Content Outline,* 2015, Planning, 6, Task 8

38. d. Identify the problem

    The first and most important is to identify the problem as a gap to be closed or as an objective to be achieved. Causes then are found by looking at the problem statement and asking why until a root cause has been identified for which action can be taken or the reasonable possibilities on the diagram have been exhausted. [Planning and Monitoring and Controlling]

    PMI®, *PMBOK® Guide*, 2013, 236, 252
    PMI® *PMP Examination Content Outline,* 2015, Monitoring and Controlling, 9, Task 1
    PMI® *PMP Examination Content Outline,* 2015, Planning, 6, Task 8

39. d. Scatter diagram

    Scatter diagrams are also called correlation charts as noted in the question. They are plot oriented pairs (X and Y). The direction of the correlation in them may be proportional (a positive correlation), inverse (a negative correlation), or a pattern of correlation may exist (a zero correlation). If a correlation can be established, then a regression line can be calculated and used to estimate how a change to the independent variable will influence the value of the dependent variable. It is one of the seven basic quality tools, a tool and technique used in Plan Quality Management and Control Quality. [Planning and Monitoring and Controlling]

    PMI®, *PMBOK® Guide*, 2013, 237, 252
    PMI® *PMP Examination Content Outline,* 2015, Monitoring and Controlling, 9, Task 1
    PMI® *PMP Examination Content Outline,* 2015, Planning, 6, Task 8

40. b. It will enable a sharper focus on the project's value proposition

Reviewing the quality management plan early ensures decisions are based on accurate information. The review also focuses more sharply on the project's value proposition, reduction in costs, and in the frequency of schedule overruns caused by rework. [Planning]

PMI®, *PMBOK® Guide*, 2013, 241
PMI® *PMP Examination Content Outline,* 2015, Planning, 6, Task 8

Project Human Resource Management

Study Hints

The Project Human Resource Management questions on the PMP® certification exam focus heavily on organizational structures, roles and responsibilities of the project manager, team building, and conflict resolution. Many of the questions are taken from the *PMBOK® Guide* and the following PMI® handbooks, which have been consolidated into one publication available from PMI® entitled *Principles of Project Management* (1997).

- *Conflict Management for Project Managers* by John R. Adams and Nicki S. Kirchof
- *Organizing for Project Management* by Dwayne P. Cable and John R. Adams
- *Roles and Responsibilities of the Project Manager* by John R. Adams and Brian W. Campbell
- *Team Building for Project Managers* by Linn C. Stuckenbruck and David Marshall
- *The Project Manager's Work Environment: Coping with Time and Stress* by Paul C. Dinsmore, Martin Dean Martin, and Gary T. Huettel

Appendix X3 *PMBOK® Guide* also should be reviewed along with six other publications it mentions:

- *Essential People Skills for Project Managers* by Ginger Levin and Steven Flannes
- *Organizing Projects for Success*, vol. 1 of *The Human Aspects of Project Management* by Vijay K. Verma
- *Human Factors in Project Management* (Revised Edition) by Paul C. Dinsmore

- *Human Resource Skills for the Project Manager*, vol. 2 of *The Human Aspects of Project Management* by Vijay K. Verma
- *Managing the Project Team*, vol. 3 of *The Human Aspects of Project Management* by Vijay K. Verma
- *Seven Habits of Highly Effective People* by Stephen R. Covey

In contrast to other areas of the *PMBOK® Guide* in which commonly known terms are used, some terminology developed for Project Human Resource Management appears to be peculiar to PMI®. (In fact, much of the terminology has been used in project management literature for many years, but that literature has not always been widely disseminated.) For example, in the area of project organizational structures, some experts with years of experience in the field have not encountered such terms or concepts as *project expeditor* or *weak matrix*. Accordingly, committing to memory PMI®'s definition and classification of the following subject areas is imperative:

- Project organizational structures
- Stages of team development
- Decision-making guidelines
- Influencing guidelines
- Negotiation skills
- Conflict management concepts

In spite of the unfamiliarity of some of the terminology, most exam takers do not find the human resource questions on the exam difficult.

PMI® views Project Human Resource Management as having four processes: Plan Human Resource Management, Acquire Project Team, Develop Project Team, and Manage Project Team. See *PMBOK® Guide*, 2013, Figure 9-1 for an overview of this structure. Know it cold!

Following is a list of the major Project Human Resource Management topics. Use it to help focus your study efforts on the areas most likely to appear on the exam.

Major Topics

Forms of organization

- Functional
- Project expeditor
- Project coordinator
- Weak matrix
- Strong matrix
- Balanced matrix
- Projectized
- Composite

Plan Human Resource Management Plan tools and techniques

- Organization chart and position descriptions
- Hierarchical-type charts
- Matrix-based charts
- Text-oriented formats
- Networking
- Organizational theory
- Expert judgment

Develop Human Resource Management Plan outputs

- Human resource management plan
- Roles and responsibilities
- Staffing management plan
- Project organization charts

Acquire Project Team
Project manager roles and responsibilities

- Functions
- Roles
- Negotiation

Types of power
Acquisition
Multi-criteria decision analysis
Virtual teams
Project staff assignments
Resource calendars
Develop Project Team objectives
Interpersonal skills

- Communication skills
- Emotional intelligence
- Conflict resolution
- Negotiation
- Influence
- Team building
- Group facilitation

Training
Team-building activities

- Approaches
- Stages of team development
- Goals and results of project team building
- Symptoms of poor teamwork
- The team-building process

Ground rules for project team building
Motivation theories

- Maslow's Hierarchy of Needs
- McGregor's Theory X and Theory Y
- Herzberg's Theory of Motivation
- Expectancy Theory
- McClelland Needs Theory

Colocation
Reward and recognition systems
Performance assessment tools
Team performance assessment

Manage Project Team

- Project conflict
- Why conflict is unavoidable on projects
- Seven sources of conflict in project environments
- Conflict and the project life cycle
- Conflict management
  - Problem solving or collaborating
  - Compromising or reconciling
  - Smoothing or accommodating
  - Withdrawal or avoiding
  - Forcing or directing

Observation and conversation
Project performance appraisals
Interpersonal skills

- Leadership
- Influencing
- Effective decision making

Political and cultural awareness
Trust building
Coaching
Change requests

Practice Questions

INSTRUCTIONS: Note the most suitable answer for each multiple-choice question in the appropriate space on the answer sheet.

1. You have been assigned as project manager on what could be a "bet the company" project. You realize that to be successful you need to exercise maximum control over project resources. Which form of project organization should you establish for this project?

   a. Strong matrix
   b. Projectized
   c. Project coordinator
   d. Weak matrix

2. Which of the following is a ground rule for project team building?

   a. Perform frequent performance appraisals
   b. Ensure that each team member reports to his or her functional manager in addition to the project manager
   c. Start early
   d. Try to solve team political problems

3. Project A is being administered using a matrix form of organization. The project manager reports to a senior vice president who provides visible support to the project. In this scenario, which of the following statements best describes the relative power of the project manager?

   a. The project manager will probably not be challenged by project stakeholders.
   b. In this strong matrix, the balance of power is shifted to the functional line managers.
   c. In this tight matrix, the balance of power is shifted to the project manager.
   d. In this strong matrix, the balance of power is shifted to the project manager.

4. You are leading a team to recommend an equitable reward and recognition system for project managers. Before finalizing the plan, you want to ensure that executives understand the basic objective of reward systems. This objective is to—

   a. Be comparable with the award system established for functional managers to indicate parity and to show the importance of project management to the company
   b. Make the link between project performance and reward clear, explicit, and achievable
   c. Motivate project managers to work toward common objectives and goals as defined by the company
   d. Attract people to join the organization's project management career path

5. Which of the following factors contributes the most to team communication?

   a. External feedback
   b. Performance appraisals
   c. Smoothing over of team conflicts by the project manager
   d. Colocation

6. You are managing a virtual team. The project has been under way for several months, and you believe your team members do not view themselves as a team or a unified group. To help rectify this situation, you should—

   a. Ensure that every member of the project team uses e-mail as a form of communication
   b. Mandate that the team follow the vision and mission statement of his or her organization
   c. Enhance communications planning
   d. Provide team members with the latest in communications technology and mandate its use

7. Major difficulties arise when multiple projects need to be managed in the functional organizational structure because of—

   a. The level of authority of the project manager
   b. Conflicts over the relative priorities of different projects in competition for limited resources
   c. Project team members who are focused on their functional specialty rather than on the project
   d. The need for the project manager to use interpersonal skills to resolve conflicts informally

8. Leadership is embedded in the job of a project manager and really in the job of every team member. Assume you are project manager with a team of three people who will work full-time on the project, and five people who will support the project on a part-time basis. All team members know one another and have worked together in the past. Leadership on this team is especially important—

   a. As it is close to the due date for deliverables
   b. At the beginning of the project
   c. During the execution phase
   d. When phase gate and performance reviews are held

9. Your organization is characterized by hierarchical organizational structures with rigid rules and policies and strict supervisory controls. Individual team members are not expected to engage in problem solving or use creative approaches to plan and execute work; management does that. Your organization is characterized by which one of the following theories?

   a. Ouchi's Theory
   b. McGregor's Theory X
   c. Maslow's self-esteem level
   d. Vroom's Expectancy Theory

10. As you prepare your human resource plan, you need to determine the skill and capacity required to complete the activities in the project. This should be documented in the—

    a. Roles and responsibilities sections
    b. Staffing management plan
    c. Staff acquisition section
    d. Competencies section

11. The primary result of effective team development is—

    a. Improved project performance
    b. An effective, smoothly running team
    c. An understanding by project team members that the project manager is ultimately responsible for project performance
    d. Enhancement of the ability of stakeholders to contribute as individuals and team members

12. The team members on your project have been complaining that they do not have any sense of identity as a team because they are located in different areas of the building. To remedy this situation, you developed a project logo and had it printed on T-shirts to promote the project, but this action has not worked. Your next step is to—

    a. Initiate a newsletter
    b. Create an air of mystery about the project
    c. Establish a "team meeting room"
    d. Issue guidelines on how team members should interact with other stakeholders

13. The project team directory is an output from which of the following processes?

    a. Develop Project Team
    b. Acquire Project Team
    c. Develop Human Resource Management Plan
    d. Manage Project Team

14. You realize that leadership without management or management without leadership probably will produce poor project results. Which one of the following key responsibilities best represents project leadership?

    a. Developing a vision and strategy and motivating people to achieve them
    b. Getting things done through other people
    c. Using charismatic power to motivate others even if they do not like the work
    d. Using all types of power, as appropriate, as motivational tools

15. Given that you are neighbors, you and the CEO of your company have established a friendly personal relationship. Recently your company appointed you project manager for a new project that is crucial to achieving next year's financial targets. Which type of power available to project managers might you be able to rely upon?

    a. Referent
    b. Reward
    c. Formal
    d. Expert

16. You have been a project manager for seven years. You now are managing the construction of a new facility that must comply with the government's newly issued environmental standards. You want to ensure that your team members are able to select methods to complete various activities on the project without needing to involve you in each situation. As you prepare your human resource management plan, you should document this information in which of the following—

    a. Roles and responsibilities section
    b. Authority section
    c. Resource breakdown structure
    d. Staffing management plan

17. It is important on all projects to determine when and how human resources will be met. Assume that you are managing a project to assess methods for streamlining the regulatory approval process for new medical devices in your government agency. Because the agency has undergone downsizing during the past three years, subject matter experts are in short supply. You must determine whether the needed subject matter experts can be acquired from inside the agency or whether you must use contractors. This information should be documented in the—

    a. Make-or-buy decisions in the procurement management plan
    b. Contracts management plan
    c. Staffing management plan
    d. Resource management plan

18. Conflicts are inevitable on projects, but if actually managed effectively, conflict can help a team arrive at a better solution. One of the biggest challenges a project manager faces is—

    a. Determining the sources of a conflict early and having proactive solutions to manage them
    b. Recognizing stakeholders will have conflicting interests but using a collaborative approach to resolve each conflict
    c. Knowing when a need for consensus is paramount
    d. Managing conflict when it occurs

19. As project manager, you are primarily responsible for implementing the project management plan by authorizing the execution of project activities. Because you do not work in a projectized organization, you do not have direct access to human resource administrative activities. Therefore you need to—
    a. Outsource these functions
    b. Prepare a project team charter that is signed off by a member of the human resources department to delineate responsibilities
    c. Ensure that your team is sufficiently aware of administrative requirements to ensure compliance
    d. Ask the head of human resources to approve your project human resource plan personally

20. Constant bickering, absenteeism, and substandard performance have characterized the behavior of certain members of your team. You have planned an off-site retreat for the team to engage in a variety of activities. Your primary objective for investing time and money in this event is to improve—
    a. Team performance
    b. Morale
    c. Quality
    d. Individual performance

21. Two team members on your project often disagree. You need a conflict resolution method that provides a long-term resolution. You decide to use which one of the following approaches?
    a. Confronting
    b. Problem solving
    c. Collaborating
    d. Smoothing

22. Which of the following is an enterprise environmental factor that may influence the development of the human resource management plan?
    a. The organizational structure of the performing organization
    b. Poor communication among team members
    c. Ambiguous staffing requirements
    d. Team morale

23. As a project manager, you believe in using a "personal touch" to further team development. One approach that has proven effective toward this goal is—
    a. Creating a team name
    b. Providing flexible work time
    c. Issuing a project charter
    d. Using conversations with team members

24. Your project has been under way for some time, and you are working in a matrix environment with only a few direct reports. However, it is your responsibility to—

    a. Facilitate an atmosphere conductive to success
    b. Focus on ensuring team member strengths are known and assign them to roles to enable the project to be completed as soon as possible
    c. Develop a recovery plan to help complete the project
    d. Provide project performance appraisals

25. You are the project manager for a two-year project that is now beginning its second year. The mix of team members has changed, and there is confusion as to roles and responsibilities. In addition, several of the completed work packages have not received the required sign-offs, and three work packages are five weeks behind schedule. To gain control of this project, you need to—

    a. Re-baseline your original human resource plan with current resource requirements
    b. Change to a projectized organizational structure for maximum control over resource assignments
    c. Work with your team to prepare a responsibility assignment matrix
    d. Create a new division of labor by assigning technical leads to the most critical activities

26. You are part of a team that is working to develop a new medical implant device. Your project manager is an expert in medical implantation devices, yet he continually seeks opinions from the team about a wide variety of project and product issues. He is also a proponent of networking in order to—

    a. Enhance project management professional development
    b. Assess the roles required for the project considering standard role descriptions in the organization
    c. Determine the most effective reporting relationships
    d. Provide guidelines on lead time for staffing based on lessons learned

27. The major difference between the project coordinator and project expeditor forms of organization is that—

    a. Strong commitment to the project usually does not exist in the project expeditor form of organization
    b. The project coordinator cannot personally make or enforce decisions
    c. The project expeditor acts only as an intermediary between management and the project team
    d. The project coordinator reports to a higher-level manager in the organization

28. Which one of the following represents a constraint on the Acquire Project Team process?

    a. Pre-assignment of staff to the project
    b. Economic factors
    c. Use of outsourcing
    d. Team member training requirements

29. Selection criteria are often used as a tool and technique in Acquire Project Team. A multi-criteria decision tool is recommended. An example of a selection criterion is—

    a. Competency
    b. Motivation
    c. Attitude
    d. Trust

30. Your team of five people to develop a database to assess testing methods for listeria in ice cream has worked together before. You expect that this team will be a high-performing team from the beginning, but you find two team members seem to disagree as to the vision of the project. You meet with these two people and find one of them really dislikes being on the team. You decide to ask your sponsor if she can be reassigned, and another person added to the team. The sponsor agrees. This means—

    a. Initial establishment of roles and responsibilities change so the new person assumes the previous team member's work
    b. Your team is in the forming stage
    c. You should conduct an off-site team-building retreat
    d. You should meet with each team member and establish specific performance goals

31. Your organization is adopting a project-based approach to business, which has been difficult. Although project teams have been created, they are little more than a collection of functional and technical experts who focus on their specialties. You are managing the company's most important project. As you begin this project, you must place a high priority on—

    a. Creating an effective team
    b. Identifying the resources needed to finish the project on time
    c. Determining the best way to communicate status to the CEO
    d. Establishing firm project requirements

32. A best practice in setting clear expectations regarding acceptable behavior by team members is to—
    a. Set ground rules
    b. Have team members sign off on the project's vision and mission statements
    c. Set up a RACI chart
    d. Link team member competencies to their specific responsibilities

33. Attitude surveys, specific assessments, structured interviews, ability tests, and focus groups are examples of—
    a. Team performance assessments
    b. Project performance appraisals
    c. Personnel assessment tools
    d. Ways to improve knowledge and skills of team members

34. Work performance reports are especially helpful in the Manage Project Team process because—
    a. They assess team performance to take corrective or preventive action as needed
    b. They assist in determining future human resource requirements
    c. They assess the degree of the project manager's authority
    d. They document and monitor who is responsible for resolving key issues and when resolution is needed

35. The key way for a project manager to promote optimum team performance in project teams whose members are not colocated is to—
    a. Build trust
    b. Establish a reward and recognition system
    c. Obtain the support of the functional managers in the other locations
    d. Exercise his or her right to control all aspects of the project

36. Hierarchical-type charts are a tool and technique for use in Plan Human Resource Management. Which one of the following is helpful in tracking project costs and can be aligned with the organization's accounting system?
    a. RACI
    b. RAM
    c. RBS
    d. OBS

37. A key benefit of the Plan Human Resource Management process is to determine the project organization chart. Therefore, an important first step is to—

    a. Create the WBS and let it determine the project organizational structure
    b. Review the project management plan
    c. Refer to the project charter developed by top management and the project sponsor
    d. Develop a project schedule, including a top-down flowchart, and identify the functional areas to perform each task

38. A resource histogram is useful as part of the human resource management plan as it—

    a. Provides a visual representation of resource allocation
    b. Enables a smooth transition to release resources
    c. Sets priorities as to when scarce resources should be scheduled
    d. Shows whether the team is colocated or virtual

39. In addition to individual performance appraisals, team performance assessments are critical. Their evaluation may include indicators such as—

    a. Reduced staff turnover rate
    b. Negotiation skills
    c. Team preferences
    d. The ability to work well with others

40. Determining the method and the timing of releasing team members should be included in the—

    a. Staff acquisition plan
    b. Human resource plan
    c. Staffing management plan
    d. Project training plan

Answer Sheet

1.	a	b	c	d	21.	a	b	c	d
2.	a	b	c	d	22.	a	b	c	d
3.	a	b	c	d	23.	a	b	c	d
4.	a	b	c	d	24.	a	b	c	d
5.	a	b	c	d	25.	a	b	c	d
6.	a	b	c	d	26.	a	b	c	d
7.	a	b	c	d	27.	a	b	c	d
8.	a	b	c	d	28.	a	b	c	d
9.	a	b	c	d	29.	a	b	c	d
10.	a	b	c	d	30.	a	b	c	d
11.	a	b	c	d	31.	a	b	c	d
12.	a	b	c	d	32.	a	b	c	d
13.	a	b	c	d	33.	a	b	c	d
14.	a	b	c	d	34.	a	b	c	d
15.	a	b	c	d	35.	a	b	c	d
16.	a	b	c	d	36.	a	b	c	d
17.	a	b	c	d	37.	a	b	c	d
18.	a	b	c	d	38.	a	b	c	d
19.	a	b	c	d	39.	a	b	c	d
20.	a	b	c	d	40.	a	b	c	d

Answer Key

1. b. Projectized

   In a projectized organizational structure, all project team members report directly and solely to the project manager. He or she has complete control over these resources and, therefore, exercises more authority over them than in any other project organizational structure. The project manager has a great amount of independence and authority. Team members are often colocated, and virtual collaboration techniques are often used to accomplish the benefits of the colocated teams. Refer to Figure 2-5 in the *PMBOK® Guide*, 2013 for an example. [Planning]

   PMI®, *PMBOK® Guide*, 2013, 22, 25
   PMI® *PMP Examination Content Outline,* 2015, Planning, 6, Task 5

2. c. Start early

   Starting the team-building process early in the project is crucial for setting the right tone and preventing bad habits and patterns from developing. Recognize that although it is essential at the front end of a project, it is ongoing. Changes in the project's environment will occur and to manage them effectively a continued or renewed team-building effort is required. It is a tool and technique in the Develop Project Team process. It can vary from a five-minute agenda item in a staff meeting to an off-site retreat designed to improve interpersonal relationships. Its objective is to help team members work together effectively. [Executing]

   Adams, J.R., et al. *Principles of Project Management,* 1997, 137
   PMI®, *PMBOK® Guide*, 2013, 276, 514
   PMI® *PMP Examination Content Outline,* 2015, Executing, 8, Task 1

3. d. In this strong matrix, the balance of power is shifted to the project manager.

   The project manager's ability to influence project decisions increases the higher up he or she—and the person to whom he or she reports—is placed in the organization. This reporting relationship in this example shows it is a strong matrix. In the strong matrix, the project manager's authority ranges from moderate to high. Therefore, it resembles some characteristics of the projectized organization because it has full-time project managers with considerable authority and full-time project administrative staff. Refer to Figure 2-4 in the *PMBOK® Guide*, 2013 for an example. [Planning]

   PMI®, *PMBOK® Guide*, 2013, 22–24
   Verma, V.K, *Organizing for Project Success,* 1995, 156–157
   PMI® *PMP Examination Content Outline,* 2015, Planning, 6, Task 5

4. b. Make the link between project performance and reward clear, explicit, and achievable

Reward and recognition systems are formal management actions that provide an incentive to behave in a particular way, usually with respect to achieving certain goals. Such systems are described in the staffing management plan. A best practice is to give the team recognition throughout the life cycle rather than waiting until the project is complete. Clear criteria for rewards and a planned system for their use help promote and reinforce desired behavior. The most effective systems are based on activities under a person's control. They are a tool and technique in the Develop Project Team process because it involves recognizing and rewarding desirable behavior. Note that a reward given to an individual can be effective only if it satisfies a need the individual values. They are typically given as part of performance appraisals in the Manage Project Team process. [Planning and Executing]

PMI®, *PMBOK® Guide,* 2013, 266, 277
PMI® *PMP Examination Content Outline,* 2015, Planning, 6, Task 5
PMI® *PMP Examination Content Outline,* 2015, Executing, 8, Task 1

5. d. Colocation

Colocation is the placement of some, many, or all of the most active team members in the same physical location to enhance their ability to perform as a team, primarily by increasing communication as to improve working relationships and productivity. It is a tool and technique in the Develop Project Team process and can be called a "tight matrix". It can be temporary or for the entire project. [Executing]

PMI®, *PMBOK® Guide,* 2013, 277, 532
PMI® *PMP Examination Content Outline,* 2015, Executing, 8, Task 1

6. c. Enhance communications planning

Because the dispersed project team does not share the same physical space each day, the possibility for misunderstandings, isolationism, difficulty in sharing information, and the cost of technology can be key issues. The project manager must enhance communications planning in the virtual team as it requires even more communication than colocated teams. Additional time also may be needed to set expectations, determine how best to resolve conflicts, involve people in making decisions, understand cultural differences, and share credit for success. [Executing]

PMI®, *PMBOK® Guide,* 2013, 271
PMI® *PMP Examination Content Outline,* 2015, Executing, 8, Task 1

7. b. Conflicts over the relative priorities of different projects in competition for limited resources

   When a finite group of resources must be distributed across multiple projects, conflicts in work assignments occur. Other sources of conflict are scheduling and personal work styles. Conflict management is a tool and technique in the Manage Project Team process. Successful conflict management results in greater productivity and more positive working relationships. The success of project managers in managing project teams often depends on their ability to resolve conflict since conflict is inevitable in the project environment. [Executing]

   PMI®, *PMBOK® Guide*, 2013, 282–283, 518
   PMI® *PMP Examination Content Outline,* 2015, Executing, 8, Task 1

8. b. At the beginning of the project

   Leadership focuses the efforts of a group of people toward a common goal to enable them to work as a team. It is known as getting work done through others. It requires respect and trust as key elements. While it is important at all phases of the project, it is crucial when a project begins, even in this situation where the people know one another and have worked together before. At the beginning, it emphasizes communicating the vision of the project and then motivating and inspiring others to achieve high performance, ensuring success. Successful projects require strong leadership skills, [Planning and Executing]

   PMI®, *PMBOK® Guide*, 2013, 284, 513
   PMI® *PMP Examination Content Outline,* 2015, Planning, 6, Task 5
   PMI® *PMP Examination Content Outline,* 2015, Executing, 8, Task 1

9. b. McGregor's Theory X

   McGregor observed two types of managers and classified them by their perceptions of workers. Theory X managers thought that workers were lazy, needed to be watched and supervised closely, and were irresponsible. Theory Y managers thought that, given the correct conditions, workers could be trusted to seek responsibility and work hard at their jobs. Ideally, the Theory X approach is not followed, but instead a flexible leadership style is used that adapts to the changes in the team's maturity throughout the project life cycle. However, it must be recognized that different organizational structures have different individual responses, individual performance results, and personal relationship characteristics. [Planning]

   McGregor, D., *The Human Side of Enterprise,* 1960, 33–35
   Verma, V.K., *Human Resource Skills for the Project Manager,* 1996, 70–71
   PMI®, *PMBOK® Guide*, 2013, 263
   PMI® *PMP Examination Content Outline,* 2015, Planning, 6, Task 5

10. d. Competency section

The competency section of this plan describes the skills and capacities required to complete activities within the project constraints. When team members do not have the required competencies, project performance may be jeopardized, and the project manager must have proactive responses to handle these situations such as training, hiring, schedule changes, or scope changes. [Planning]

PMI®, *PMBOK® Guide*, 2013, 264
PMI® *PMP Examination Content Outline,* 2015, Planning, 6, Task 5

11. a. Improved project performance

Improved project performance not only increases the likelihood of meeting project objectives, it also creates a positive team experience contributing to the enhancement of team capabilities. It results in improved teamwork, enhanced people skills and competencies, motivated employees, reduced staff turnover rates, and improved overall team performance. It is a benefit of the Develop Project Team process. [Executing]

PMI®, *PMBOK® Guide*, 2013, 273
PMI® *PMP Examination Content Outline,* 2015, Executing, 8, Task 1

12. c. Establish a "team meeting room"

Colocating team members, even on a temporary basis, enhances communications, thereby contributing to improved project performance. In addition, the team meeting room (often called a "war room") provides a sense of identity to the project team and raises the visibility of the project within the organization. This room can serve as a place to post schedules and other conveniences that enhance communication. If a virtual team is used, a virtual team room or war room can be set up if the technology is available. [Executing]

PMI®, *PMBOK® Guide*, 2013, 277
PMI® *PMP Examination Content Outline,* 2015, Executing, 8, Task 1

13. b. Acquire Project Team

The project team directory is part of project staff assignments, an output from the Acquire Project Team process. Within staff assignments, memos to project team members and names inserted into other parts of the project management plan such as the project organization chart and schedules are included. Other outputs are resource calendars and updates to the project management plan. [Executing]

PMI®, *PMBOK® Guide*, 2013, 272
PMI® *PMP Examination Content Outline,* 2015, Executing, 8, Task 1

14. d. Developing a vision and strategy and motivating people to achieve them

    Leadership involves developing a vision of the future and strategies to achieve that vision, positioning people to carry out the vision, and helping people energize themselves to overcome any barriers to change and achieve high performance. Throughout the project, leaders are responsible for establishing and maintaining this vision, strategy, and communicating about it. It is important to lead by example and follow through with commitments. [Executing]

    PMI®, *PMBOK® Guide*, 2013, 284, 513–515
    PMI® *PMP Examination Content Outline,* 2015, Executing, 8, Task 1

15. a. Referent

    Referent power is based on a less powerful person's identification with a more powerful person. This type of power is useful in terms of persuasion and helps the project manager exert influence over individuals from whom he or she needs support. French and Bell, in 1973, formalized the five sources of power. In addition, to referent as described in this answer, reward is power in which the project manager can give team members rewards and recognition, leading to this section as a tool and technique in Develop Project Team. Coercive power is a leader who makes it clear that if team members do not comply with the organization's or the leader's mandate, there will be repercussions; comparable to McGregor in the discussion of Theory X. Legitimate power is shown by a project manager that he or she has a legitimate right to influence others and expects others to follow his or her approach in the project; in a way it supports influencing skills, a tool and technique in Manage Project Team. Expert power is when one uses his or her experience or knowledge to influence others, again supporting influencing skills. By recognizing the project manager and other leader's sources of power, the team member then can consider appropriate ways to best work with the project manager for project success, As noted in the *PMBOK® Guide*, it is necessary to apply power skillfully and cautiously to ensure long-term collaboration. [Planning]

    Adams, J.R., et al., *Principles of Project Management,* 1997, 174–180
    Levin, G., *Interpersonal Skills for Portfolio, Program, and Project Managers,* 2010, 162–167
    PMI®, *PMBOK® Guide*, 2013, 277, 284, 515
    PMI® *PMP Examination Content Outline,* 2015, Executing, 8, Task 1

16. b. Authority section

Authority refers to the right to apply project resources, make decisions, and sign approvals. Examples include selecting methods to complete activities, quality acceptance, and responding to variances in the project. The individual authority of each team member should match their individual responsibilities; when this is done it is easier for the team members to do their work. This is documented in the authority section in the human resource management plan. [Planning]

PMI®, *PMBOK® Guide*, 2013, 264
PMI® *PMP Examination Content Outline,* 2015, Planning, 6, Task 5

17. c. Staffing management plan

The staffing management plan is part of the human resource management plan. One section of it involves staff acquisition. Among other things, this section includes whether the human resources will come from within the organization or from external, contracted sources. These data then help to plan the acquisition of project team members. This section also includes whether the team members will be colocated or virtual, the cost associated for each level of expertise needed for the project, and the level of assistance the human resources department and the functional managers can provide to the project team. [Planning]

PMI®, *PMBOK® Guide*, 2013, 265
PMI® *PMP Examination Content Outline,* 2015, Planning, 6, Task 5

18. d. Managing conflict when it occurs

In doing so the project manager draws on the other interpersonal skills to lead the team to a successful resolution of the situation. This means project managers must develop skills and have the experience to adapt their personal conflict management style to the situation. It also is necessary to focus on building trust in the project environment and by doing so to engage people to find a positive resolution to the situation. Conflict management is a tool and technique in the Manage Project Team process. [Executing]

PMI®, *PMBOK® Guide*, 2013, 282–283
PMI® *PMP Examination Content Outline,* 2015, Executing, 8, Task 1

19. c. Ensure that your team is sufficiently aware of administrative requirements to ensure compliance

A projectized work environment is unusual because project managers rarely have every function under their control. However, compliance is a section in the project's staffing management plan, which is part of the project's human resource management plan. This section includes strategies for complying with any applicable government regulations, union contracts, or other established human resource policies, leading to this question where the key words are "direct access to human resource administrative activities". By including this section, the team can best recognize what is necessary in this environment. [Planning]

PMI®, *PMBOK® Guide*, 2013, 22, 267
PMI® *PMP Examination Content Outline,* 2015, Planning, 6, Task 5

20. a. Team performance

Team development leads to improved team performance, which ultimately results in improved project performance. Improvements in team performance can come from many sources and can affect many areas of project performance. For example, improved individual skill levels, such as enhanced technical competence, may enable team members to perform their assigned activities more effectively. Team development efforts have greater benefit when conducted early but should take place throughout the project life cycle. Team building activities are a tool and technique in the Develop Project Team process and are considered an ongoing process, crucial to project success. Team performance assessments are an output of this project in which team building is noted as an activity that can increase team performance, as well as the likelihood of meeting project objectives. [Executing]

PMI®, *PMBOK® Guide*, 2013, 274, 278
PMI® *PMP Examination Content Outline,* 2015, Executing, 8, Task 1

21. c. Collaborating

Collaborating or problem solving is an effective technique for managing conflict when a project is too important to be compromised. It involves incorporating multiple insights and viewpoints from people with different perspectives and offers a good opportunity to learn from others. It provides a long-term resolution. It requires a cooperative attitude and open dialogue that typically leads to consensus and commitment. [Executing]

PMI®, *PMBOK® Guide*, 2013, 283, 518
Verma, V.K., *Human Resource Skills for the Project Manager,* 1996, 119–120
PMI® *PMP Examination Content Outline,* 2015, Executing, 8, Task 1

22. a. The organizational structure of the performing organization

    Enterprise environmental factors can influence the Develop Human Resource Management process. The organizational structure of the performing organization determines whether the project manager's role is a strong one (as in a strong matrix) or a weak one (as in a weak matrix). Other examples of enterprise environmental factors are the organization's culture, geographic dispersion of team members, existing human resources, personnel administration functions, and marketplace conditions. [Planning]

    PMI®, *PMBOK® Guide*, 2013, 260
    PMI® *PMP Examination Content Outline*, 2015, Planning, 6, Task 5

23. d. Using conversations with team members

    Project managers can show interest in their team members by communicating with them regularly. Observation of their work and conversations with them are effective approaches to stay in touch with them and gauge their attitudes. Both observation and conversation, a tool and technique in Manage Project Team, also enables a way to monitor project performance, accomplishments that are a source of pride to the team members, and any interpersonal issues so proactive actions can be taken to best resolve them. [Executing]

    PMI®, *PMBOK® Guide*, 2013, 282
    PMI® *PMP Examination Content Outline*, 2015, Executing, 8, Task 1

24. d. Provide project performance appraisals

    As a tool and technique in Manage Project Team, performance appraisals can clarify roles and responsibilities, provide feedback to team members, discover and resolve any unknown or unresolved issues, develop individual training plans to increase competencies, and establish specific goals for future time periods. The need for formal or informal performance appraisals depends on the complexity of the project and its length, as in this question. [Executing]

    PMI®, *PMBOK® Guide*, 2013, 282
    PMI® *PMP Examination Content Outline*, 2015, Executing, 8, Task 1

25. c. Work with your team to prepare a responsibility assignment matrix

The responsibility assignment matrix (RAM) defines project roles and responsibilities in terms of work packages and activities. It can be used to show who is a participant, who is accountable, who handles review, who provides input, and who must sign off on specific work packages or project phases. RAMs can be at different levels. A high-level RAM defines what the project team is responsible for within each WBS component, and lower-level RAMs are used within the team as described above in this rationale. It ensures there is only one person accountable for any single task to avoid confusion over responsibilities. An example of a RAM is a RACI (responsible, accountable, consult, and inform) chart, as shown in Figure 9-5 in the *PMBOK® Guide*, 2013. While these matrix-based charts are an input to Plan Human Resource Management, a best practice is to review them regularly, especially as new people join the project and others leave or if there are changes to the work packages and their associated activities. [Planning]

PMI®, *PMBOK® Guide*, 2013, 262, 559
PMI® *PMP Examination Content Outline,* 2015, Planning, 6, Task 5

26. a. Enhance project management professional development

Networking is an input to Plan Human Resource Management and is the formal and informal interaction with others in the organization, industry, or professional environment. It is a constructive way to understand political and interpersonal factors that may impact staffing management decisions. It is a useful technique at the beginning of the project and also can be an effective way to enhance project management professional development during the project and even after the project ends. The other answers to this question refer to the use of expert judgment in developing the human resource management plan. [Planning]

PMI®, *PMBOK® Guide*, 2013, 263
PMI® *PMP Examination Content Outline,* 2015, Planning, 6, Task 5

27. d. The project coordinator reports to a higher-level manager in the organization

The relative position of the project coordinator in the organization is thought to lead to an increased level of authority and responsibility. The project coordinator also tends to have more power to make decisions and has some authority. The project expediter cannot personally make or enforce decisions. [Executing]

Adams, J.R., et al., *Principles of Project Management,* 1997, 15–17
Verma, V.K., *Organizing Projects for Success,* 1995, 153–156
PMI®, *PMBOK® Guide*, 2013, 23
PMI® *PMP Examination Content Outline,* 2015, Executing, 8, Task 1

28. b. Economic factors

During the process of Acquire Project Team, if human resources are not available because of constraints, the project manager may be required to assign alternative resources. Examples of constraints are economic factors or previous assignments to other projects. The project manager then may need to assign people who perhaps have lower competencies, provided there is no violation of legal, regulatory, mandatory, or other specific criteria. [Executing]

PMI®, *PMBOK® Guide*, 2013, 268
PMI® *PMP Examination Content Outline,* 2015, Executing, 8, Task 1

29. c. Attitude

Attitude is used to determine one's ability to work with others as a cohesive team. Other selection criteria are availability, cost, experience, ability, knowledge, skills, and international factors. A multi-criteria approach is useful as the criteria then are weighted according to the importance of the needs within the team. [Executing]

PMI®, *PMBOK® Guide*, 2013, 271–272
PMI® *PMP Examination Content Outline,* 2015, Executing, 8, Task 1

30. b. Your team is in the forming stage

Even if team members have worked together previously, it cannot be assumed they will immediately be a high-performing team. According to the Tuchman team development model, all teams have a forming stage in which the team meets and learns about the project and their formal roles and responsibilities. At this time, team members are independent and not as open in this phase. This model further goes on to the storming stage, norming stage, performing stage, and adjourning stage. However, even if the team is in the norming or performing stage, if a team member leaves or a new person joins, the team will revert back to the forming stage. This model is part of team-building activities, a tool and technique in the Develop Project Team process. [Executing]

PMI®, *PMBOK® Guide*, 2013, 276
PMI® *PMP Examination Content Outline,* 2015, Executing, 8, Task 1

31. a. Creating an effective team

An effective team is critical to project success, but such a team is not born spontaneously. In early project phases, it is vitally important for the project manager to place a high priority on initiating and implementing the team-building process. It is the purpose of the Acquire Project Team process, which has as its key benefit outlining and guiding the team selection process and resource assignment to obtain a successful team. Insufficient resources or capabilities decrease the probability of success and may even led to project termination. [Executing]

PMI®, *PMBOK® Guide*, 2013, 267–268
Verma, V.K., *Managing the Project Team,* 1997, 137
PMI® *PMP Examination Content Outline,* 2015, Executing, 8, Task 1

32. a. Set ground rules

Ground rules are a tool and technique in the Develop Project Team process. Early commitment to clear guidelines decreases misunderstandings and increases productivity. Discussing how they are to address such areas as a code of conduct, communications, working together, or meeting etiquette are examples of setting ground rules. This allows team members to discover values that are important to one another. Once the ground rules are set, all project team members have the responsibility to see they are followed. [Executing]

PMBOK® Guide, 2013, 277
PMI® *PMP Examination Content Outline,* 2015, Executing, 8, Task 1

33. c. Personnel assessment tools

Personnel assessment tools are a tool and technique in the Develop Project Team process. They are used to give the project manager and the team members' insight into areas of strength and areas needing improvement. They help the project manager assess team members' preferences and aspirations, how they process and organize information, how they make decisions, and how they prefer to interact with others. They can help provide improved understanding, trust, commitment, and communications among team members, leading to a more productive team. [Executing]

PMBOK® Guide, 2013, 278
PMI® *PMP Examination Content Outline,* 2015, Executing, 8, Task 1

34. b. They assist in determining future human resource requirements

Work performance reports are an input in the Manage Project Team process. They provide information on the current project status compared to forecasts. Performance areas that can help with team management include results from schedule, cost, and quality control and scope validation. The information from these reports then assists in determining future human resource requirements, recognition and rewards, and updates to the staffing management plan. [Executing]

PMI®, *PMBOK® Guide*, 2013, 281
PMI® *PMP Examination Content Outline,* 2015, Executing, 8, Task 1

35. a. Build trust

Whether a team is colocated or virtual, it is essential to build trust for effective team leadership. Without trust, it is difficult to establish the positive relationships in the team, and if trust is compromised, relationships deteriorate, people disengage, and collaboration may be impossible. It is particularly essential in the virtual team in which team members who are physically separate from one another tend not to know each other well. They have few opportunities to develop trust in the traditional way, and they tend to communicate poorly with one another. Trust then must become the foundation upon which all team-building activities are built. Without trust, it is hard to even know if a virtual team member is working productively on the project or is doing other work, knowledge transfer becomes more difficult if not impossible, and communications planning is hindered. Virtual teams are a tool and technique in the Acquire Project Team process. [Executing]

PMI®, *PMBOK® Guide*, 2013, 271, 517–518
PMI® *PMP Examination Content Outline,* 2015, Executing, 8, Task 1

36. c. RBS

While the WBS shows how project deliverables are broken down into work packages to show high-level areas of responsibility, the organizational breakdown structure (OBS) is arranged according to the organization's departments, units, or teams with the work packages listed under each one as appropriate. The resource breakdown structure (RBS) is a hierarchical list of resources related by category and resource type used to facilitate and control the work of the project. Each descending lower level represents an increasing detailed description of the resource unit small enough to be used in conjunction with the WBS to enable the work to be planned, monitored, and controlled. The RBS helps track project costs and can be aligned with the organization's accounting system. It contains all resources, not solely people. [Planning]

PMI®, *PMBOK® Guide*, 2013, 261
PMI® *PMP Examination Content Outline,* 2015, Planning, 6, Task 5

37. b. Review the project management plan

Other benefits of this process are to establish project roles and responsibilities and prepare the staffing management plan for the timetable for staff acquisition and release. The project management plan is the first input in this process to develop the human resource management plan. It contains key information that shows the project life cycle and the processes applied to each phase, how the work will be done to accomplish objectives, a change management and a configuration management plan, how integrity of the project baselines will be maintained, and needs and methods of communication. [Planning]

PMI®, *PMBOK® Guide*, 2013, 258–259
PMI® *PMP Examination Content Outline,* 2015, Planning, 6, Task 5

38. a. Provides a visual representation of resource allocation

The resource histogram is contained in the resource calendars section of the staffing management plan, which is part of the human resource management plan. It illustrates the number of hours a resource will be needed each week or month during a project. It can include a horizontal line that represents the maximum number of hours available for a particular resource. Bars that extend beyond this line show the need for a resource optimization strategy such as adding more resources or extending the schedule. Refer to Figure 9-6 in the *PMBOK® Guide*, 2013 for an example. [Planning]

PMI®, *PMBOK® Guide*, 2013, 265
PMI® *PMP Examination Content Outline,* 2015, Planning, 6, Task 5

39. a. Reduced staff turnover rate

Team performance assessments are an output of the Develop Product Team process. In addition to the answer, other indicators of the team's effectiveness include: improvements in skills that allow individuals to perform assignments more effectively, improvements in competencies so the team performs better as a team, and increased team cohesiveness enabling team members to transfer knowledge and openly and help one another improve overall project performance. [Executing]

PMI®, *PMBOK® Guide*, 2013, 278–279
PMI® *PMP Examination Content Outline,* 2015, Executing, 8, Task 1

40. c. Staffing management plan

The staffing management plan is a document that describes when and how human resources will become part of the project team and when they will return to their organizational units. It addresses how staff members will be acquired, how long they will remain on the project, how and when they will be released, training needs, and other important aspects of forming and disbanding the team. When team members are released from a project, the costs associated with them are no longer charged to the project, thus reducing project costs. Morale is improved when there is a smooth transition to another project or one in progress. The staff release plan helps mitigate human resource risks that may occur at the end of the project. [Planning]

PMI®, *PMBOK® Guide*, 2013, 266
PMI® *PMP Examination Content Outline,* 2015, Planning, 6, Task 5

Project Communications Management

Study Hints

The Project Communications Management questions on the PMP® certification exam are relatively basic and are taken primarily from the *PMBOK® Guide* and other PMI®-published reference materials. Common sense and your own experience will play a large role in your ability to answer the questions on this topic. There will be questions that test your specific knowledge of *PMBOK® Guide* terms and concepts. However, there will also be many general questions that require you to choose the "best" answer.

The questions focus on formal and informal communication, verbal versus written communication, performance reporting, and management styles. PMI® considers management style to be an essential component of how a project manager communicates.

The PMI® handbooks (which are now included in *Principles of Project Management*, PMI®, 1997), *Roles and Responsibilities of the Project Manager* by John R. Adams and Brian W. Campbell, *Conflict Management for Project Managers* by John R. Adams and Nicki S. Kirchof, and *Team Building for Project Managers* by Linn C. Stuckenbruck and David Marshall, should be studied thoroughly for this section of the PMP® certification exam. Also review Appendix X3 in the *PMBOK® Guide*.

The PMI® publication *Human Resource Skills for the Project Manager*, which is Volume 2 of *The Human Aspects of Project Management* by Vijay K. Verma, is another useful reference. In these publications, many mention the importance of the kickoff meeting one of the most effective mechanisms in Project Communications Management. The nature and purpose of this meeting are discussed in *Team Building for Project Managers* but it is not discussed in the *PMBOK® Guide*. However, it is Task 12 in Planning in the ECO.

PMI® views Project Communications Management as a process consisting of three elements: plan communications management, manage communications, and control communications. See *PMBOK® Guide* Figure 10-1 for an overview of this structure. Know this chart thoroughly.

Following is a list of the major Project Communications Management topics. Use it to help focus your study efforts on the areas most likely to appear on the exam.

Major Topics

Importance of project communications management
Communication dimensions
Communication channels
Communication skills

Plan communications

- The communications model
- Encode
- Transmit message
- Decode
- Acknowledge
- Feedback/response
- Communications requirements analysis
- Communications technology
- Communication methods
- Meetings
- Communications management plan

Manage communications

- Sender-feedback model
- Choice of media
- Writing style
- Meeting management techniques
- Presentation techniques
- Facilitation techniques
- Listening techniques
- Information management systems
- Performance reporting
- Simple
- Elaborate
- Project communications

Barriers to communication
Control communications

- Issue log
- Meetings
- Work performance information
- Change requests

Practice Questions

INSTRUCTIONS: Note the most suitable answer for each multiple-choice question in the appropriate space on the answer sheet.

1. The communications management process goes beyond distributing relevant information as it—
   a. Focuses on establishing working relationships and standard formats for global communication among stakeholders
   b. Ensures information being communicated has been appropriately generated, received, and understood
   c. Establishes individual and group responsibilities and accountabilities for communication management
   d. Discusses how sensitive communications are to be handled

2. One purpose of the communications management plan is to provide information about the—
   a. Methods that will be used to convey information
   b. Methods that will be used for releasing team members from the project when they are no longer needed
   c. Project organization and stakeholder responsibility relationships
   d. Experience and skill levels of each team member

3. Project managers for international projects should recognize key issues in cross-cultural settings and place special emphasis on—
   a. Establishing a performance reporting system
   b. Using good communication planning
   c. Establishing and following a production schedule for information distribution to avoid responding to requests for information between scheduled communications
   d. Using translation services for formal, written project reports

4. You are managing a project with team members located at customer sites on three different continents. As you plan communications with your stakeholders, you should review—
   a. Stakeholder management plan
   b. Stakeholder register
   c. Communications model
   d. Communications channels

5. Having worked previously on projects as a team member, you are pleased to now be the project manager to develop a new process to ensure that software projects in your IT department are considered a success and are not late or over budget. However, many of your team members are new to the organization. As you work to establish a high-performing team, you realize the importance of—

   a. Mentoring
   b. Coaching
   c. Moving quickly through the forming and storming stages
   d. Safeguarding information

6. As a project manager, you try to use empathic listening skills to help understand another person's frame of reference. In following this approach, you should—

   a. Mimic the content of the message
   b. Probe, then evaluate the content
   c. Evaluate the content, then advise
   d. Rephrase the content and reflect the feeling

7. Statements of organizational policies and philosophies, position descriptions, and constraints are examples of—

   a. Formal communication
   b. Lateral communication
   c. External communication
   d. Horizontal communication

8. You have decided to organize a study group of other project managers in your organization to help prepare for the PMP® exam. What type of communication activity are you employing in your efforts to organize this group?

   a. Horizontal
   b. Vertical
   c. Official
   d. External

9. Your CEO just sent you an e-mail asking you to make a presentation on your project, which has been in progress for 18 months, to over 50 identified internal and external stakeholders. You have been conducting such presentations and holding meetings regularly on this important project. You should begin by—

   a. Defining the audience
   b. Determining the objective
   c. Deciding on the general form of the presentation
   d. Circulating issues to be discussed

10. You are responsible for a project in your organization that has multiple internal customers. Because many people in your organization are interested in this project, you realize the importance of—

    a. Conducting a stakeholder analysis to assess information needs
    b. Performing communications planning early
    c. Determining the communications requirements of the customers
    d. Having an expert on communications management and customer relationship management on your team

11. Project managers spend a great deal of time communicating with the team, the stakeholders, the client, and the sponsor. One can easily see the challenges involved, especially if one team member must communicate a technical concept to another team member in a different country. The first step in this process is to—

    a. Encode the message
    b. Decode the message
    c. Determine the feedback loops
    d. Determine the medium

12. On your project, scope changes, constraints, assumptions, integration and interface requirements, and overlapping roles and responsibilities pose communications challenges. The presence of communication barriers is most likely to lead to—

    a. Reduced productivity
    b. Increased hostility
    c. Low morale
    d. Increased conflict

13. The most common communication problem that occurs during negotiation is that—

    a. Each side may misinterpret what the other side has said
    b. Each side may give up on the other side
    c. One side may try to confuse the other side
    d. One side may be too busy thinking about what to say next to hear what is being said

14. You finally have been appointed project manager for a major company project. One of your first activities as project manager will be to create the communications management plan. As you match the stakeholder with the appropriate communication methods for that stakeholder, you could use any one of the following methods EXCEPT—

    a. Interactive communications
    b. Passive communications
    c. Pull communications
    d. Push communications

15. As an output of Plan Communications, it may be necessary to update the project documents, which include the—

    a. Stakeholder register
    b. Corporate policies, procedures, and processes
    c. Knowledge management system
    d. Stakeholder management plan

16. Sample attributes of a communications management plan include which one of the following?

    a. Roles
    b. Responsibilities
    c. Ethics
    d. Authority

17. A skill in negotiating is—

    a. Active listening
    b. Attentive listening
    c. Getting to "yes"
    d. Using confrontation to solve problems

18. The key benefit of the Control Communications process is—

    a. Sharing best practices with other project teams in the organization with lessons learned
    b. Ensuring the information needs of stakeholders are met
    c. Ensuring an optimal information flow among communication participants
    d. Providing stakeholders with information about resolved issues, approved status, and project status

19. The issue log is useful in Control Communications because it—
    a. Provides what has happened and is a platform for subsequent communications
    b. Includes the project's risk register
    c. Organizes and summarizes information gathered
    d. Serves as an information management system for communications management

20. As head of the PMO, you will receive performance reports for all major projects. You decided to set a guideline for project managers as performance reporting should—
    a. Collect work performance information on the status of deliverables
    b. Provide earned value data for project forecasting
    c. Provide information at an appropriate level for each audience
    d. Focus on cost and schedule variances rather than scope, resources, quality, and risks

21. A simple performance report provides information on—
    a. Percent complete
    b. Customer satisfaction
    c. Unacceptable variances
    d. Scope creep

22. Communication is important when setting and managing expectations with the stakeholders. Which one of the following statements is NOT true regarding the importance of communications within a project?
    a. Communications is one of the single biggest contributors to project success or failure.
    b. Project resources should be spent primarily on communicating information that leads to project success.
    c. Effective communications includes awareness of communication styles, cultural issues, relationships, personalities, and the context of the situation
    d. Listening is part of communicating and is a way to gain insight into problem areas, managing conflicts, and making decisions.

23. In person-to-person communication, messages are sent on verbal levels and nonverbal levels simultaneously. To communicate effectively, the project manager should be—
    a. Aware of the communication styles of the other parties
    b. Aware of his or her own preferred communication style
    c. Able to apply a flexible style to meet audience requirements
    d. Aware of non-verbal communications

24. As an output from Control Communications, it may be necessary to update the—
    a. Identified risks
    b. Forecasts
    c. Corporate policies, procedures, and processes
    d. Knowledge management system

25. In project communications, the first step in a written communication is to—
    a. Analyze the facts and assumptions that have a bearing on the purpose of the message
    b. Gather thoughts or ideas
    c. Develop a logical sequence of the topics to be addressed
    d. Establish the basic purpose of the message

26. A communications management plan includes which one of the following sample contents?
    a. Issues
    b. Escalation processes, including time frames and the management chains
    c. Dimensions
    d. Project assumptions and constraints

27. Your organization has decided to use project management for all of its endeavors. It has established a Center of Excellence for Project Management to support the movement into management by projects and has appointed you as its director. Since you work in a matrix environment, which of the following types of communications is the most essential for success?
    a. Upward
    b. Horizontal
    c. Downward
    d. Diagonal

28. You have heard recently that the client calls your progress reports the "Code of Hammurabi" because they seem to be written in hieroglyphics and are completely indecipherable to all but an antiquities scholar. This situation could have been avoided by—
    a. Informing the client at the start of the project about the types of reports they will receive
    b. Using risk management techniques to identify client issues
    c. Hiring an expert report writer to prepare standard reports
    d. Engaging in communications planning

29. Assume on your project you have identified 250 stakeholders located in three continents and of these 250, you have determined that 200 of them will be actively involved and interested in your project. Therefore, as you determine an appropriate communication method, your best approach is—

    a. Elaborate status reports
    b. Simple status reports
    c. Knowledge repositories
    d. E-mails

30. You want to ensure that the information you collect showing project progress and status is meaningful to stakeholders. You want to combine the type and format of the stakeholder's information needs with an analysis of the value of the information. You will document this information in the—

    a. Communications register
    b. Stakeholder register
    c. Stakeholder management plan
    d. Communications management plan

31. Work performance information is an output of which process?

    a. Manage risks
    b. Manage communications
    c. Control communications
    d. Report performance

32. Assume you want to optimize the work performance reports you will use to manage communications. You should do so by—

    a. Determining the most appropriate choice of communications media
    b. Setting different communications techniques for different stakeholder groups
    c. Ensuring the information is consistent with regulations and standards
    d. Ensuring comprehensiveness, accuracy, and availability

33. Information received from stakeholders concerning project operations can be distributed and used to modify or improve future performance of the project. This modification or improvement is done as an update to organizational process assets during which of the following processes?

    a. Plan communications management
    b. Distribute information
    c. Manage communications
    d. Report performance

34. General management skills relevant to the Manage Communications process include—
    a. Operational planning
    b. Organizational behavior
    c. Setting and managing expectations
    d. Influencing the organization

35. Changes in the report formats and lessons learned documents process should trigger changes to the—
    a. Project management plan and performance reporting system
    b. Integrated change control system and the communications management plan
    c. Monitor and control project process and the project management plan
    d. Organizational process assets updates

36. One way to determine how to best update and communicate project performance and respond to stakeholder information requests is to—
    a. Review the effectiveness of the communications management plan
    b. Set up a portal
    c. Hold meetings
    d. Distribute performance reports

37. The purpose of work performance data in Control Communications is to present results of comparative analysis to the—
    a. Performance measurement baseline
    b. Communications management plan
    c. Stakeholder management plan
    d. Deliverable status

38. Because communications planning often is linked tightly with enterprise environmental factors, which one of the following statements is true?
    a. The project's organizational structure has a major effect on the project's communications requirements.
    b. Standardized guidelines, work instructions, and performance measurement criteria are key items to consider.
    c. Procedures for approving and issuing work authorizations should be taken into consideration.
    d. Criteria and guidelines to tailor standard processes to the specific needs of the project should be stated explicitly.

39. You are working on a project with 15 stakeholders. The number of communication channels on this project is—

    a. 15
    b. 105
    c. 210
    d. 225

40. Which of the following formulas calculates the number of communication channels in a project?

    a. $\frac{n(n-1)}{2}$
    b. $\frac{n^2-1}{2}$
    c. $\frac{n^2-1}{n}$
    d. $\frac{2^n-2}{1^n}$

Answer Sheet

1.	a	b	c	d	21.	a	b	c	d
2.	a	b	c	d	22.	a	b	c	d
3.	a	b	c	d	23.	a	b	c	d
4.	a	b	c	d	24.	a	b	c	d
5.	a	b	c	d	25.	a	b	c	d
6.	a	b	c	d	26.	a	b	c	d
7.	a	b	c	d	27.	a	b	c	d
8.	a	b	c	d	28.	a	b	c	d
9.	a	b	c	d	29.	a	b	c	d
10.	a	b	c	d	30.	a	b	c	d
11.	a	b	c	d	31.	a	b	c	d
12.	a	b	c	d	32.	a	b	c	d
13.	a	b	c	d	33.	a	b	c	d
14.	a	b	c	d	34.	a	b	c	d
15.	a	b	c	d	35.	a	b	c	d
16.	a	b	c	d	36.	a	b	c	d
17.	a	b	c	d	37.	a	b	c	d
18.	a	b	c	d	38.	a	b	c	d
19.	a	b	c	d	39.	a	b	c	d
20.	a	b	c	d	40.	a	b	c	d

Answer Key

1. b. Ensures information being communicated has been appropriately generated, received, and understood

   In addition the Manage Communications process focuses on providing opportunities for stakeholders to request further information, clarification, and discussion. It enables effective and efficient information to flow between stakeholders. [Executing]

   PMI®, *PMBOK® Guide*, 2013, 297–298
   PMI® *PMP Examination Content Outline,* 2015, Executing, 8, Task 6

2. a. Methods that will be used to convey information

   These methods or technologies can include memos, e-mails, and press conferences. They are one of several items to include in this plan. This plan, an output of the Plan Communications Management process, describes how project communications will be planned, structured, monitored, and controlled. [Planning]

   PMI®, *PMBOK® Guide*, 2013, 296
   PMI® *PMP Examination Content Outline,* 2015, Planning, 6, Task 6

3. b. Using good communications planning

   An effective way to manage cultural diversity on projects is for the project manager to get to know the team members and to use good communication planning. It is necessary to consider time zones and language barriers, as well as cross-cultural differences, and to include a glossary of common terminology in the communications management plan. [Planning]

   PMI®, *PMBOK® Guide*, 2013, 290, 296, 516
   PMI® *PMP Examination Content Outline,* 2015, Planning, 6, Task 6

4. b. Stakeholder register

   The stakeholder register is an input to the Plan Communications Management process. It contains the identified stakeholders including their name, position, location, and role; their main requirements, expectations, and potential influence; and whether or not they are supporters, neutral, or resistors of the project. [Planning]

   PMI®, *PMBOK® Guide*, 2013, 291, 398
   PMI® *PMP Examination Content Outline,* 2015, Planning, 6, Task 6

5. b. Coaching

Many communications skills are common to both general management and project management. Coaching is one example. It is especially useful to develop the team to higher levels of competency and performance and helping people recognize their potential through empowerment and development. It is used to aid team members to develop or enhance their skills required to achieve project success. It can be a powerful motivator for teams and can lead to greater efficiency and productivity. [Planning and Executing]

PMI®, *PMBOK® Guide*, 2013, 288, 519
PMI® *PMP Examination Content Outline,* 2015, Planning, 6, Task 6
PMI® *PMP Examination Content Outline,* 2015, Executing, 8, Task 6

6. d. Rephrase the content and reflect the feeling

Empathic listening requires seeing the world the way the other person sees it, with the goal of understanding that person's views and feelings. Unlike sympathetic listening, empathic listening contains no element of value judgment. It is essential to listen actively and effectively and to question and probe ideas to help ensure better understanding. Active listening and passive listening give the user insight into problem areas, negotiation and conflict resolution strategies, decision making, and problem solving. [Executing]

Covey, S.R., *The 7 Habits of Highly Effective People: Powerful Lessons in Personal Change,* 2013, 229–235
PMI®, *PMBOK® Guide*, 2013, 288, 515
PMI® *PMP Examination Content Outline,* 2015, Executing, 8, Task 6

7. a. Formal communication

Formal communication provides direction and control for project team members and other employees. They also contain reports, minutes, and briefings and are examples of organizational process assets used in Plan Communications Management, Manage Communications and in Control Communications. [Planning, Executing and Monitoring and Controlling]

PMI®, *PMBOK® Guide*, 2013, 287, 300, 306
PMI® *PMP Examination Content Outline,* 2015, Planning, 6, Task 6
PMI® *PMP Examination Content Outline,* 2015, Executing, 8, Task 6
PMI® *PMP Examination Content Outline,* 2015, Monitoring and Controlling, 9, Task 1

8. a. Horizontal

   Communication activities have many potential dimensions to consider in exchanging information between the sender and the receiver. Horizontal communication occurs between or among peers, that is, across, rather than up and down, the organization. [Executing]

   PMI®, *PMBOK® Guide*, 2013, 287
   PMI® *PMP Examination Content Outline,* 2015, Executing, 8, Task 6

9. d. Circulating issues to be discussed

   Meetings are held regularly on projects to update and communicate project information and to respond to requests from stakeholders for the information. Most meetings consist of stakeholders coming together to resolve problems or make decisions. Typical meetings begin with a defined list of issues to be discussed, which are distributed in advance with minutes and other key information relative to the meeting. Such meetings are a tool and technique in the Plan Communications Management process. [Planning]

   PMI®, *PMBOK® Guide*, 2013, 295
   PMI® *PMP Examination Content Outline,* 2015, Planning, 6, Task 6

10. b. Performing communications planning early

    On most projects, communications planning should be performed very early such as when the project management plan is prepared. This approach then allows appropriate resources, such as time and budget, to be allocated to communications activities. [Planning]

    PMI®, *PMBOK® Guide*, 2013, 290
    PMI® *PMP Examination Content Outline,* 2015, Planning, 6, Task 6

11. a. Encode the message

    As the first step in the basic communication model, it is essential to translate thoughts or ideas into a language that is understood by others. Then, the message is sent using various technologies, and the receiver decodes it or translates it back into meaningful thoughts or ideas, acknowledges it, and may provide feedback and a response. It is a tool and technique in the Plan Communications Management process. [Planning]

    PMI®, *PMBOK® Guide*, 2013, 293
    PMI® *PMP Examination Content Outline,* 2015, Planning, 6, Task 6

12. d. Increased conflict

Barriers to communication lead to a poor flow of information. Accordingly, messages are misinterpreted by recipients, thereby creating different perceptions, understanding, and frames of reference. Left unchecked, poor communication increases conflict among project stakeholders, which causes the other problems listed to arise. Then, the project manager must work actively to resolve conflicts so disruptive impacts are prevented. [Executing]

PMI®, *PMBOK® Guide*, 2013, 288, 300
Verma, V.K., *Managing the Project Team,* 1997, 24–25
PMI® *PMP Examination Content Outline,* 2015, Executing, 8, Task 6

13. a. Each side may misinterpret what the other side has said

Effective communication is the key to successful negotiation, which is a key communication skill. Misunderstanding is the most common communication problem. A project manager should listen actively, acknowledge what is being said, and speak for a purpose. It is essential to listen attentively and communicate articulately. Negotiation is an integral part of project management and if done well increases the probability of project success. [Executing]

PMI®, *PMBOK® Guide*, 2013, 288, 517
Fisher, R., Ury, W., and Patton, B. *Getting to Yes: Negotiation Without Giving In*, 2nd ed., 1991, 32–34
Verma, V.K., *Human Resource Skills for the Project Manager*, 1996, 165
PMI® *PMP Examination Content Outline,* 2015, Executing, 8, Task 6

14. b. Passive communications

You can use several different methods to share information. Interactive communications are multi-directional in nature, such as conferences and meetings. Pull communications are those methods where the recipients find the information at their leisure and get the information that they want at their discretion. Push communications is targeted information sent to a select group but does not certify that the recipient actually has received the information, such as e-mail. Passive communications is more of a style of delivering the content or receiving the content. [Planning]

PMI®, *PMBOK® Guide*, 2013, 295
PMI® *PMP Examination Content Outline,* 2015, Planning, 6, Task 6

15. a. Stakeholder register

In the Plan Communications Management process the two documents that may be updated are the project schedule and the stakeholder register. The stakeholder register is an input to this process but may be changed, as the communications management plan is prepared based on communications requirements analysis. This analysis determines the information needs of the project stakeholders. [Planning]

PMI®, *PMBOK® Guide*, 2013, 291, 297
PMI® *PMP Examination Content Outline*, 2015, Planning, 6, Task 6

16. b. Responsibilities

Among other things, the communications management plan should identify the person responsible for communicating the information and the person responsible for authorizing release of any confidential information. [Planning]

PMI®, *PMBOK® Guide*, 2013, 296
PMI® *PMP Examination Content Outline*, 2015, Planning, 6, Task 6

17. b. Attentive listening

Negotiation if done well increases the probability of project success and involves conferring with others of shared or opposed interests with a view toward compromise. Negotiating is required to achieve mutually acceptable agreements between parties. A key skill in negotiating is the ability to both listen attentively and communicate articulately. [Executing]

PMI®, *PMBOK® Guide*, 2013, 288, 517
PMI® *PMP Examination Content Outline*, 2015, Executing, 8, Task 6

18. b. Ensuring optimal information flow among all communication participants

While control communications as a process monitors and controls communications throughout the project to ensure the communication needs of project stakeholders are met, the key benefit is to ensure an optimal information flow among all communication participants at any moment in time. This process can trigger the iteration of the Plan Communications Management and Manage Communications processes based on some key issues and the impact of possible repercussions of project communications. [Monitoring and Controlling]

PMI®, *PMBOK® Guide*, 2013, 303–304
PMI® *PMP Examination Content Outline*, 2015, Monitoring and Controlling, 9, Task 1

19. a. Provides what has happened and is a platform for subsequent communications

    The issue log is an input to Control Communications and is used to document and monitor issue resolution. It can facilitate communications and ensure a common understanding of issues. In this process its information provides a repository of what already has happened in the project and serves as a platform for subsequent communications to be delivered. It documents and helps to monitor who is responsible for resolving certain issues by a target date. It also addresses obstacles that can block the team from achieving its goals. [Monitoring and Controlling]

    PMI®, *PMBOK® Guide*, 2013, 305
    PMI® *PMP Examination Content Outline,* 2015, Monitoring and Controlling, 9, Task 5

20. c. Provide information at an appropriate level for each audience

    Performance reporting is a tool and technique in Manage Communications. Performance reports range from simple status reports to more elaborate reports. The emphasis is to ensure performance reporting provides the needed information for each audience level. It involves the periodic collection and analysis of performance versus actual data to understand and communicate project progress and to forecast project results. [Executing]

    PMI®, *PMBOK® Guide*, 2013, 301
    PMI® *PMP Examination Content Outline,* 2015, Executing, 8, Task 6

21. a. Percent complete

    A simple status report may show performance information such as percent complete or status information for each area (scope, schedule, cost, and quality). [Executing]

    PMI®, *PMBOK® Guide*, 2013, 301
    PMI® *PMP Examination Content Outline,* 2015, Executing, 8, Task 6

22. b. Project resources should be spent primarily on communicating information that leads to project success

    Communications is considered one of the single most powerful indicators of project success or failure. Effective communications includes an awareness of all types of filters that may be impeding or straining communications. Listening is vital to good communications. Resources also should be spent on determining where a lack of communications can lead to failure. Planning project communications is important to the ultimate success of any project as inadequate communications planning may lead to problems including message delivery delays, communicating information to the wrong people, and insufficient communications to stakeholders, which can lead to misunderstandings or misinterpretations. [Planning]

    PMI®, *PMBOK® Guide*, 2013, 289–291
    PMI® *PMP Examination Content Outline,* 2015, Planning, 6, Task 6

23. a. Aware of the communication styles of the other parties

    It also is important to consider nuances and norms, relationships, personalities, and the situation. These actions lead to mutual understanding and thus to effective communications. Other techniques for effective communications include sender/receiver models, choice of media, writing style, meeting management techniques, presentation techniques, facilitation techniques, and listening techniques. [Executing]

    PMI®, *PMBOK® Guide*, 2013, 298–299, 515
    PMI® *PMP Examination Content Outline,* 2015, Executing, 8, Task 6

24. b. Forecasts

    Communications control often entails the need to update project documents, including forecasts, performance reports, and the issue log. While corporate policies, procedures, and processes are typical organizational process assets, in this process, the organizational process assets to update are report formats and lessons learned documentation. [Monitoring and Controlling]

    PMI®, *PMBOK® Guide*, 2013, 308
    PMI® *PMP Examination Content Outline,* 2015, Monitoring and Controlling, 9, Task 1

25. b. Gather thoughts or ideas

For any type of communication, the first step in the basic communication model is to encode, which means the sender translates thoughts or ideas into language. It is followed by transmitting the message, decoding it by the receiver, acknowledging its receipt from the receiver, and providing feedback and responses and then transmitting them back to the original sender. Communication models are a tool and technique in the Plan Communications Management process. [Planning]

PMI®, *PMBOK® Guide*, 2013, 293
PMI® *PMP Examination Content Outline,* 2015, Planning, 6, Task 6

26. b. Escalation processes, including time frames and the management chains

Numerous items, including escalation processes, are part of the communications management plan. Business issues may arise that cannot be resolved at a lower staff level. During such a time, an escalation process is required to show time frames and the names of people in the management chain who will work to resolve these issues. [Planning]

PMI®, *PMBOK® Guide*, 2013, 296
PMI® *PMP Examination Content Outline,* 2015, Planning, 6, Task 6

27. b. Horizontal

Horizontal communication is between the project manager and his or her peers and will be where most of the communications will occur. Accordingly, it is essential for success in a highly competitive environment and requires diplomacy, experience, and mutual respect. [Executing]

Verma, V.K., *Managing the Project Team*, 1997, 136
PMI®, *PMBOK® Guide*, 2013, 287
PMI® *PMP Examination Content Outline,* 2015, Executing, 8, Task 6

28. d. Engaging in communications planning

    The communications management plan is prepared during Plan Communications Management. The plan should include a description of the information to be distributed such as format, content, level of detail, as well as conventions and definitions to be used. It also can include templates and guidelines for meetings and e-mails. In preparing this plan, and especially in this scenario in the question, meetings are a useful tool and technique. They can be used, among other things, to determine the most appropriate way to update and communicate project information and to respond to stakeholders. By having client representatives as part of such meetings, the problems in this scenario may have been prevented. [Planning]

    PMI®, *PMBOK® Guide*, 2013, 296–297
    PMI® *PMP Examination Content Outline,* 2015, Planning, 6, Task 6

29. c. Knowledge repositories

    Knowledge repositories along with Intranet sites, e-learning, and lessons learned data bases are examples of methods of pull communications. They are used for large volumes of information or for large audiences and require recipients to access communication content at their own discretion. They are a communication method, a tool and technique in the Plan Communications Management process. [Planning]

    PMI®, *PMBOK® Guide*, 2013, 295
    PMI® *PMP Examination Content Outline,* 2015, Planning, 6, Task 6

30. d. Communications management plan

    The project team must conduct an analysis of stakeholder communications requirements to ensure that stakeholders are receiving the information required to participate in the project. For example, stakeholders typically require performance reports for information purposes. Such information requirements should be included in the communications management plan. Conducting a communications requirements analysis to determine the stakeholders' information needs is a tool and technique in the Plan Communications process. [Planning]

    PMI®, *PMBOK® Guide*, 2013, 291, 296
    PMI® *PMP Examination Content Outline,* 2015, Planning, 6, Task 6

31. c. Control Communications

    Work performance information, an output of Control Communications, organizes and summaries performance data such as status and progress information on the project at the level required by stakeholders. This information next is communicated to the appropriate stakeholders. [Monitoring and Controlling]

    PMI®, *PMBOK® Guide*, 2013, 307
    PMI® *PMP Examination Content Outline,* 2015, Monitoring and Controlling, 9, Task 1

32. d. Ensuring comprehensiveness, accuracy, and availability

    Work performance reports are an input to Manage Communications. They are a collection of project performance and status information used to facilitate discussion and create communications. They should be comprehensive, accurate, and available in a timely way. [Executing]

    PMI®, *PMBOK® Guide*, 2013, 299
    PMI® *PMP Examination Content Outline,* 2015, Executing, 8, Task 6

33. c. Manage Communications

    Feedback from stakeholders is an example of an organizational process asset to update as a result of the Manage Communications process. It used to modify or improve future project performance. Other organizational process assets to update are stakeholder notifications, project reports, project presentations, project records, and lessons learned documentation. [Executing]

    PMI®, *PMBOK® Guide*, 2013, 302–303
    PMI® *PMP Examination Content Outline,* 2015, Executing, 8, Task 6

34. c. Setting and managing expectations

    Communications skills are part of general management skills, and setting and managing expectations are an example in Manage Communications. Setting expectations helps create, collect, distribute, store, retrieve, and ultimately dispose of project information according to the communications management plan. By doing so it enables an efficient and effective communications flow among project stakeholders. [Executing]

    PMI®, *PMBOK® Guide*, 2013, 287–288, 297
    PMI® *PMP Examination Content Outline,* 2015, Executing, 8, Task 6

35. d. Organizational process assets updates

Any changes in report formats and lessons learned documentation are organizational process asset updates as an output of Control Communications. The documentation may become part of the historical database for both the project and the organization and may include the causes of issues, reasons behind the corrective actions taken, and other lessons learned in the project. [Monitoring and Controlling]

PMI®, *PMBOK® Guide*, 2013, 308
PMI® *PMP Examination Content Outline*, 2015, Monitoring and Controlling, 9, Task 6

36. c. Hold meetings

Meetings are a tool and technique in Control Communications. They can be face to face or online and in different locations and may include not only the project team but also suppliers, vendors, and other stakeholders. It is important to hold these meetings to better determine the most appropriate way to update and communicate project performance and to respond to requests from stakeholders for information. [Monitoring and Controlling]

PMI®, *PMBOK® Guide*, 2013, 307
PMI® *PMP Examination Content Outline*, 2015, Monitoring and Controlling, 9, Task 1

37. a. Performance measurement baseline

Work performance data are an input in Control Communications. These data organize and summarize information gathered and present the results of comparative analysis to the performance measurement baseline. [Monitoring and Controlling]

PMI®, *PMBOK® Guide*, 2013, 305
PMI® *PMP Examination Content Outline*, 2015, Monitoring and Controlling, 9, Task 1

38. a. The project's organizational structure has a major effect on the project's communications requirements.

Enterprise environmental factors undoubtedly will influence the project's success and must be considered because communication must be adapted to the project environment. They are an input to the Plan Communications Management Plan process. [Planning]

PMI®, *PMBOK® Guide*, 2013, 29, 291
PMI® *PMP Examination Content Outline*, 2015, Planning, 6, Task 6

39. b. 105

The formula for determining the number of communication channels is n(n – 1)/2, where n = the number of stakeholders: 15(15 – 1)/2 = (15)(14)/2 = 105. It is important to note that project managers must plan the project's communications requirements carefully, limiting who will communicate with whom given the potential for confusion when multiple communications channels can exist. Determining these communications channels is part of communications requirements analysis, a tool and technique in the Plan Communications Management process. [Planning]

PMI®, *PMBOK® Guide*, 2013, 291–292
PMI® *PMP Examination Content Outline,* 2015, Planning, 6, Task 6

40. a. $\frac{n(n-1)}{2}$

Where n = the number of stakeholders. This formula, as noted in the previous rationale, is important to learn as you may see a test question on it in the exam. It is necessary to determine the communications channels to help determine the needs of the project's stakeholders and to ensure resources are spent on communicating only the information that contributes to project success or where a lack of communications can lead to failure. [Planning]

PMI®, *PMBOK® Guide*, 2013, 292
PMI® *PMP Examination Content Outline,* 2015, Planning, 6, Task 6

Project Risk Management

Study Hints

Most exam takers find the Project Risk Management questions on the PMP® certification exam demanding because they address many concepts that project managers may not have been exposed to in their work or education. However, the questions correspond closely to *PMBOK® Guide* material, so you should not have much difficulty if you study the concepts and terminology found there. Although the questions included do not contain mathematically complex work problems, they do require you to know certain theories, such as expected monetary value (EMV) and decision-tree analysis. Additionally, you are likely to encounter questions related to levels of risk faced by both buyer and seller based on various types of contracts.

PMI® views risk management as a six-step process including Plan Risk Management, Identify Risks, Perform Qualitative Risk Analysis, Perform Quantitative Risk Analysis, Plan Risk Responses, and Control Risk. *PMBOK® Guide* Figure 11-1 provides an overview of this approach. Know this chart thoroughly.

Another useful reference is PMI's *Practice Standard for Risk Management,* which you can read on the PMI web site.

Following is a list of the major Project Risk Management topics. Use it to help focus your study efforts on the areas most likely to appear on the exam.

Major Topics

Project risk management

- Risk defined
- Types of risk
- Known risks
- Unknown risks
- Organization's risk attitude
- Risk appetite
- Risk tolerance
- Risk threshold
- Risk factors
- Risk event
- Probability of occurrence
- Amount at stake (impact)
- Risk conditions
- Risk tolerances

Risk processes
Plan Risk Management

- Planning meetings and analyses
- Analytical techniques
- Risk management plan
- Methodology
- Roles and responsibilities
- Budget
- Timing
- Categories
- Definitions of probability and impact
- Probability and impact matrix
- Revised stakeholder tolerances
- Reporting formats
- Tracking

Identify Risks

- Definition
- Timing
- Plans and baselines
- Project documents

Identify Risks tools and techniques

- Documentation reviews
- Brainstorming
- Delphi method
- Interviews
- Root-cause analysis
- Strengths-weaknesses-opportunities-threats (SWOT) analysis
- Influence diagrams
- Checklists
- Assumption analysis
- Diagramming techniques
- Expert judgment

Risk register

- List of identified risks
- List of potential responses

Perform Qualitative Risk Analysis

- Prioritize risks for further action
- Risk probability and impact assessment
- Probability and impact matrix
- Risk data quality assessment
- Risk categories
- Risk urgency assessment
- Expert judgment
- Assumption log updates
- Risk register updates
- Relative ranking or priority list of risks
- Risks by category
- Causes of risks requiring particular attention
- Risks requiring near-term responses and additional analysis and responses
- Watch list of low priority risks

Perform Quantitative Risk Analysis

- Numerical analysis of the effect of identified risks on project objectives
- Interviewing
- Probability distribution
- Sensitivity analysis
- Expected monetary value analysis
- Decision-tree analysis

- Monte Carlo analysis
- Path convergence
- Statistical distribution
- Risk register updates
- Probabilistic analysis of the project
- Probability of achieving cost and time objectives
- Prioritized list of quantified risks
- Trends

Plan Risk Responses

- Negative risks or threats
  - Avoid
  - Transfer
  - Mitigate
  - Accept
- Positive risks or opportunities
  - Exploit
  - Share
  - Enhance
  - Accept
- Contingent responses
- Risk register updates
- Risk-related contract decisions
- Updates to plans and documents

Control Risks

- Definition and purpose
- Tools and techniques
- Risk reassessment
- Risk audits
- Variance and trend analysis
- Technical performance measurement
- Reserve analysis
- Status meetings
- Updates to the risk register
- Change requests
- Updates to organizational process assets, plans, and documents

Practice Questions

INSTRUCTIONS: Note the most suitable answer for each multiple-choice question in the appropriate space on the answer sheet.

1. As the project manager, you have the option of proposing one of three systems to a client: a full-feature system that not only satisfies the minimum requirements but also offers numerous special functions (the "Mercedes"); a system that meets the client's minimum requirements (the "Yugo"); and a system that satisfies the minimum requirements plus has a few extra features (the "Toyota"). The on-time records and associated profits and losses are depicted on the below decision tree. What is the expected monetary value of the "Toyota" system?

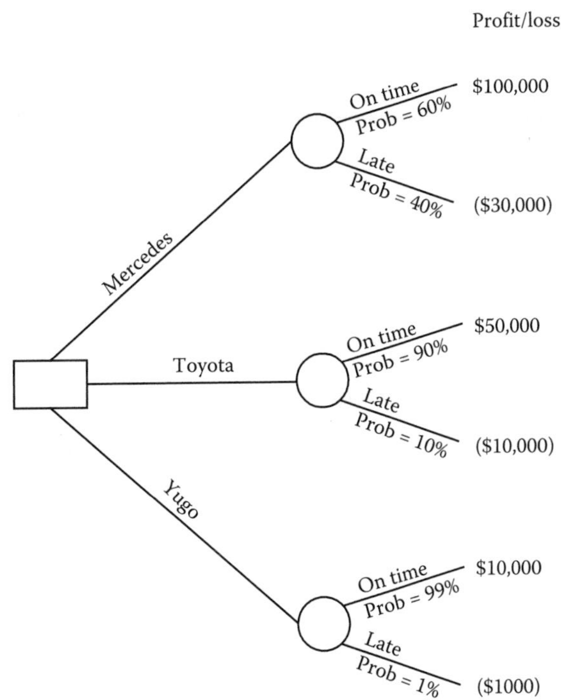

   a. $9,900
   b. $44,000
   c. $45,000
   d. $48,000

2. A risk response strategy that can be used for both threats and opportunities is—

   a. Share
   b. Avoid
   c. Accept
   d. Transfer

3. The risk urgency assessment is a tool and technique used for—
   a. Plan Risk Responses
   b. Identify Risks
   c. Perform Qualitative Risk Analysis
   d. Perform Quantitative Risk Analysis

4. Projects are particularly susceptible to risk because—
   a. Murphy's law states that "if something can go wrong, it will"
   b. There is uncertainty in all projects
   c. Project management tools are generally unavailable at the project team level
   d. There are never enough resources to do the job

5. As project manager, you have assembled the team to prepare a comprehensive list of project risks. Which one of the following documents would be the most helpful in this process?
   a. OBS
   b. WBS
   c. RBS
   d. CBS

6. You are working on identifying possible risks to your project to develop a nutritional supplement. You want to develop a comprehensive list of risks that can be addressed later through qualitative and quantitative risk analysis. An information gathering technique used to identify risks is—
   a. Documentation reviews
   b. Probability and impact analysis
   c. Checklist analysis
   d. Brainstorming

7. The Delphi technique is a particularly useful method for identifying risks to—
   a. Present a sequence of decision choices graphically to decision makers
   b. Define the probability of occurrence of specific variables
   c. Reduce bias in the analysis and keep any one person from having undue influence on the outcome
   d. Help take into account the attitude of the decision maker toward risk

8. A workaround is—

   a. An unplanned response to a negative risk event
   b. A plan of action to follow when something unexpected occurs
   c. A specific response to certain types of risk as described in the risk management plan
   d. A proactive, planned method of responding to risks

9. Most statistical simulations of budgets, schedules, and resource allocations use which one of the following approaches?

   a. PERT
   b. Decision-tree analysis
   c. Present value analysis
   d. Monte Carlo analysis

10. In the below path convergence example, if the odds of completing activities 1, 2, and 3 on time are 50 percent, 50 percent, and 50 percent, what are the chances of starting activity 4 on day 6?

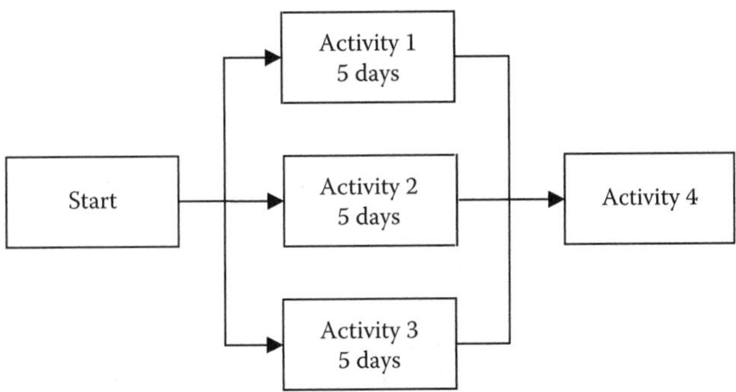

   a. 10 percent
   b. 13 percent
   c. 40 percent
   d. 50 percent

11. A project health check identified a risk that your project would not be completed on time. As a result, you are quantifying the project's risk exposure and determining what cost and schedule contingency reserves might be needed. You performed a schedule risk analysis using Monte Carlo analysis. The basis for your schedule risk analysis is the—

   a. WBS
   b. Gantt chart
   c. Schedule network diagram and duration estimates
   d. Probability/impact risk rating matrix

12. You are developing radio frequency (RF) technology that will improve overnight package delivery. You ask each stakeholder to estimate the most optimistic package delivery time using the RF technology, the most pessimistic time, and the most likely time. This shows that for your next step you plan to—

    a. Use a beta or triangular probability distribution
    b. Conduct a sensitivity analysis
    c. Structure a decision analysis as a decision tree
    d. Determine the strategy for risk response

13. Each one of the following statements about risk avoidance is true EXCEPT that it—

    a. Focuses on changing the project management plan to eliminate entirely the threat
    b. Isolates the project's objectives from the risk's impact
    c. Accepts the consequences of the risk event should it occur
    d. Changes the project objective that is in jeopardy

14. If the probability of event 1 is 80 percent and of event 2 is 70 percent and they are independent events, how likely is it that both events will occur?

    a. 6 percent
    b. 15 percent
    c. 24 percent
    d. 56 percent

15. The project scope statement should be used in the Identify Risk process because it—

    a. Identifies project assumptions
    b. Identifies all the work that must be done and, therefore, includes all the risks on the project
    c. Helps to organize all the work that must be done on the project
    d. Contains information on risks from prior projects

16. Your project team has identified all the risks on the project and has categorized them as high, medium, and low. The "low" risks are placed on which one of the following for monitoring?

    a. Threat list
    b. Low risk list
    c. Watch list
    d. Low impact list

17. A management reserve is used for—
    a. Risks that are identified at the outset of the project
    b. Risks that are not identified at the outset of the project but are known before they occur
    c. Risks that cannot be known before they occur because they are external risks
    d. Any risks that cannot be known before they occur

18. The simplest form of quantitative risk analysis and modeling techniques is—
    a. Probability analysis
    b. Sensitivity analysis
    c. Delphi technique
    d. Utility theory

19. If a business venture has a 60-percent chance to earn $2 million and a 20-percent chance to lose $1.5 million, what is the expected monetary value of the venture?
    a. –$50,000
    b. $300,000
    c. $500,000
    d. $900,000

20. You are managing the construction of a highly sophisticated data center in Port Moresby, Papua, New Guinea. Although this location offers significant economic advantages, the threat of typhoons has caused you to create a backup plan to operate in Manila in case the center is flooded. This plan is an example of what type of risk response?
    a. Passive avoidance
    b. Mitigation
    c. Active acceptance
    d. Deflection

21. A recent earned value analysis shows that your project is 20 percent complete, the CPI is 0.67, and the SPI is 0.87. In this situation, you should—
    a. Perform additional resource planning, add resources, and use overtime as needed to accomplish the same amount of budgeted work
    b. Re-baseline the schedule, then use Monte Carlo analysis
    c. Conduct a risk response audit to help control risk
    d. Forecast potential deviation of the project at completion from cost and schedule targets

22. The purpose of a numeric scale in risk management is to—
    a. Avoid high-impact risks
    b. Assign a relative value to the impact on project objectives if the risk in question occurs
    c. Rank order of risks in terms of very low, low, moderate, high, and very high
    d. Test project assumptions

23. Risk score measures the—
    a. Variability of the estimate
    b. Product of the probability and impact of the risk
    c. Range of schedule and cost outcomes
    d. Reduced monetary value of the risk event

24. Which of the following is an example of recommended corrective action in risk management?
    a. Conducting a risk audit
    b. Engaging in additional risk response planning
    c. Performing the contingency plan
    d. Conducting a risk review

25. The primary advantage of using decision-tree analysis in project risk management is that it—
    a. Considers the attitude of the decision maker toward risk
    b. Forces consideration of the probability of each outcome
    c. Helps to identify and postulate risk scenarios for the project
    d. Shows how risks can occur in combination

26. Your project is using complex, unproven technology. Your team conducted a brainstorming session to identify risks. Poor allocation of project resources was the number one risk. This risk was placed on the risk register, which included at this point a—
    a. Watch list
    b. Potential risk response
    c. Known unknown
    d. List of other risks requiring additional analysis

27. When managing current projects, it is important to use lessons learned from previous projects to improve the organization's project management process. Therefore, in the Identify Risks process one can review the—

    a. Process flow charts
    b. Checklists
    c. WBS
    d. Root-cause analysis

28. Risk mitigation involves—

    a. Using performance and payment bonds
    b. Eliminating a specific threat by eliminating the cause
    c. Avoiding the schedule risk inherent in the project
    d. Reducing the probability and/or impact of an adverse risk event to an acceptable threshold

29. On a typical project, when are risks highest and the cost of changes the lowest?

    a. During the concept phase
    b. At or near completion of the project
    c. During the implementation phase
    d. When the project manager is replaced

30. Two key inputs to the Perform Quantitative Risk Analysis process are the—

    a. WBS and milestone list
    b. Scope management plan and process improvement plan
    c. Schedule management plan and cost management plan
    d. Procurement management plan and quality baseline

31. In preparing the risk management plan meetings, the key purpose of these meetings is to—

    a. Involve those involved in executing the plan
    b. Enable anyone in the organization to participate
    c. Define a high-level plan
    d. Ensure the key stakeholders are invited and participate

32. Which one of the following statements best characterizes an activity cost or duration estimate developed with a limited amount of information?

    a. It should be part of the planning for the needed management reserve.
    b. It is an input to Identify Risks.
    c. It is an output from Identify Risks.
    d. It must be factored into the list of prioritized project risks.

33. What is the primary difference between a risk audit and a risk reassessment?

    a. A risk reassessment is conducted at the completion of a major phase; audits are conducted after the project is complete.
    b. Project stakeholders conduct risk audits; management conducts reassessments.
    c. Risk reassessments are regularly scheduled; risk audits are performed as defined in the project's risk management plan.
    d. There is no difference; they are virtually the same.

34. Accurate and unbiased data are essential for Perform Qualitative Risk Analysis. Which one of the following should you use to examine the extent of understanding of project risk?

    a. Data quality assessment
    b. Project assumptions testing
    c. Sensitivity analysis
    d. Influence diagrams

35. Assigning more talented resources to the project to reduce time to completion or to provide better quality than originally planned are examples of which one of the following strategies?

    a. Enhance
    b. Exploit
    c. Share
    d. Contingent response

36. Which of the following is NOT an objective of a risk audit?

    a. Confirming that risk management has been practiced throughout the project life cycle
    b. Confirming that the project is well managed and that the risks are being controlled
    c. Evaluating the effectiveness of risk responses in dealing with identified risks
    d. Ensuring that each risk identified and deemed critical has a computed expected value

37. Contingency planning involves—

    a. Defining the steps to be taken if an identified risk event should occur
    b. Establishing a management reserve to cover unplanned expenditures
    c. Preparing a stand-alone document that is separate from the overall project plan
    d. Determining needed adjustments to make during the implementation phase of a project

38. Assume that you are working on a new product for your firm. Your CEO learned that a competitor was about to launch a new product that has similar features to those of your project. The competitor plans to launch the product on September 1. It is now March 1. Your schedule called for you to launch your product on December 1. Your CEO now has now mandated that you fast track your project so you can launch your product on August 1. This fast track schedule is an example of an—
    a. Unknown risk
    b. A risk taken to achieve a reward
    c. A response that requires sharing the risk
    d. A passive avoidance strategy

39. As head of the project management office, you need to focus on those items where risk responses can lead to better project outcomes. One way to help you make these decisions is to—
    a. Use a probability and impact matrix
    b. Assess trends in Perform Quantitative Risk Analysis results
    c. Prioritize risks and conditions
    d. Assess trends in Perform Qualitative Risk Analysis results

40. You are the project manager for the construction of an incinerator to burn refuse. Local residents and environmental groups are opposed to this project. Management agrees to move this project to a different location. This is an example of which one of the following risk responses?
    a. Passive acceptance
    b. Active acceptance
    c. Mitigation
    d. Avoidance

Answer Sheet

1.	a	b	c	d	21.	a	b	c	d
2.	a	b	c	d	22.	a	b	c	d
3.	a	b	c	d	23.	a	b	c	d
4.	a	b	c	d	24.	a	b	c	d
5.	a	b	c	d	25.	a	b	c	d
6.	a	b	c	d	26.	a	b	c	d
7.	a	b	c	d	27.	a	b	c	d
8.	a	b	c	d	28.	a	b	c	d
9.	a	b	c	d	29.	a	b	c	d
10.	a	b	c	d	30.	a	b	c	d
11.	a	b	c	d	31.	a	b	c	d
12.	a	b	c	d	32.	a	b	c	d
13.	a	b	c	d	33.	a	b	c	d
14.	a	b	c	d	34.	a	b	c	d
15.	a	b	c	d	35.	a	b	c	d
16.	a	b	c	d	36.	a	b	c	d
17.	a	b	c	d	37.	a	b	c	d
18.	a	b	c	d	38.	a	b	c	d
19.	a	b	c	d	39.	a	b	c	d
20.	a	b	c	d	40.	a	b	c	d

Answer Key

1. b. $44,000

$$\begin{aligned} EMV_{\text{Toyota}} &= (\$50{,}000 \times 90\%) + (-\$10{,}000 \times 10\%) \\ &= \$45{,}000 + (-\$1{,}000) \\ &= \$44{,}000 \end{aligned}$$

Expected monetary value analysis (EVM) is a statistical concept that calculates the average outcome when the future scenarios may not happen, which is considered analysis under uncertainty. The EVM opportunities are generally evaluated as positive values, as in this example, while threats are expressed as negative values. Review Figure 11-16 in the *PMBOK® Guide*, 2013 for another example. [Planning]

PMI®, *PMBOK® Guide*, 2013, 339
PMI® *PMP Examination Content Outline,* 2015, Planning, 6, Task 10

2. c. Accept

Risk exists on every project, and it is unrealistic to think it can be eliminated completely. There are certain risks that simply must be accepted because we cannot control whether or not they will occur (for example, an earthquake). Acceptance is a strategy for dealing with risk that can be used for both threats and opportunities. Using it, the project team decides to acknowledge the risk and not take any action unless the risk occurs if the risk is a threat. If the risk is an opportunity, the project team will take advantage of it if it occurs but will not actively pursue it. [Planning]

PMI®, *PMBOK® Guide*, 2013, 345–346
PMI® *PMP Examination Content Outline,* 2015, Planning, 6, Task 10

3. c. Perform Qualitative Risk Analysis

Risks that may happen in the near-term need urgent attention. The purpose of the risk urgency assessment is to identify those risks that have a high likelihood of happening sooner rather than later. Indicators of priority may include the probability of detecting the risk, time to affect a risk response, symptoms and warning signs, and the risk ranking. It often is combined with the risk ranking to give a final risk severity ranking. [Planning]

PMI®, *PMBOK® Guide*, 2013, 333
PMI® *PMP Examination Content Outline,* 2015, Planning, 6, Task 10

4. b. There is uncertainty in all projects

Every project has uncertainty associated with it because a project by its definition is a temporary endeavor undertaken to create a unique product, service, or result. Risks may be known or unknown. Known risks are those that have been identified and analyzed, so it is possible to plan risk responses. Known risks then can be managed proactively using a contingency reserve. Unknown risks are ones that cannot be managed proactively and may be assigned a management reserve; often, they are considered issues. [Planning and Monitoring and Controlling]

PMI®, *PMBOK® Guide*, 2013, 3 and 310
PMI® *PMP Examination Content Outline,* 2015, Planning, 6, Task 10
PMI® *PMP Examination Content Outline,* 2015, Monitoring and Controlling, 9, Task 4

5. c. RBS

The risk breakdown structure (RBS) helps to provide a framework for ensuring a comprehensive process of systematically identified risks. It is a hierarchically organized depiction of the identified risks by risk categories. It helps the project team to look at many sources from which a risk may arise in a risk identification exercise. Different RBS structures are appropriate for different types of projects, or the organization can use a previously prepared framework that can be a simple list of categories or it may be structured into a RBS. It is described in the risk categories section of the risk management plan. It also is used as part of risk categorization to determine the areas of the project affected to determine the areas exposed to the effects of uncertainty. It is a tool and technique in Perform Quantitative Risk Analysis. [Planning]

PMI®, *PMBOK® Guide*, 2013, 317, 332
PMI® *PMP Examination Content Outline,* 2015, Planning, 6, Task 10

6. d. Brainstorming

Brainstorming is a frequently used information-gathering technique for identifying risk, because it enables the project team to develop a list of potential risks relatively quickly. Project team members, or invited experts added possibly under the direction of a facilitator, participate in the session. Traditional brainstorming sessions are free form. Categories of risks, such as in a RBS, can be used as a framework. The identified risks can be categorized by the type of risk, and then their definitions can be refined. [Planning]

PMI®, *PMBOK® Guide*, 2013, 324
PMI® *PMP Examination Content Outline,* 2015, Planning, 6, Task 10

7. c. Reduce bias in the analysis and keep any one person from having undue influence on the outcome

   The Delphi technique provides a means for arriving at a consensus using a panel of experts to determine a solution to a specific problem. Project risk experts are identified but participate anonymously. Each panelist answers a questionnaire. Then the responses, along with opinions and justifications, are summarized and are re-circulated to the experts for further comment. Consensus tends to be reached in a few rounds in this process. [Planning]

   PMI®, *PMBOK® Guide*, 2013, 324
   Wideman, M.R., *Project and Program Risk Management: A Guide to Managing Project Risks and Opportunities*, 1992, C-2 and C-3
   PMI® *PMP Examination Content Outline,* 2015, Planning, 6, Task 10

8. a. An unplanned response to a negative risk event

   Used in Control Risks, a workaround is a response to a threat that has occurred for which a prior response had not been planned or was not effective. They also are used to deal with emerging risks that were not identified or were passively accepted. It is an example of recommended corrective action and is typically handled through a change request, processed in the Integrated Change Control process. [Monitoring and Controlling]

   PMI®, *PMBOK® Guide*, 2013, 353, 567
   PMI® *PMP Examination Content Outline,* 2015, Monitoring and Controlling, 9, Task 4

9. d. Monte Carlo analysis

   A project simulation uses a model that translates the specified detailed uncertainties into their potential outcomes on project objectives and is an tool and technique in Perform Quantitative Risk Analysis, Simulations are typically performed using Monte Carlo in which a project model is computed many times with the input values (e.g., cost estimates or activity durations) chosen at random for each iteration from the probability distribution of these variables. Monte Carlo analysis supports various statistical distributions (normal, triangular, beta, uniform, etc.) used in estimating budgets, schedules, and resource allocations. See Figure 11-17 in the *PMBOK® Guide*, 2013 for an example of the results in a cost risk simulation. [Planning]

   Frame, J.D., *The New Project Management: Tools for an Age of Rapid Change, Corporate Reengineering, and Other Business Realities* 2002, 89
   PMI®, *PMBOK® Guide*, 2013, 340
   PMI® *PMP Examination Content Outline,* 2015, Planning, 6, Task 10

10. b. 13 percent

$$\text{Probability (starting activity 4 on day 6)} = (0.5)^3 = 0.125 \text{ or } 13\%$$

Such an approach helps to prioritize risks for further quantitative analysis and to plan risk responses based on their risk rating. The rating is based on their assessed probability and impact on an objective if it does occur. The probability and impact matrix is a tool and technique in the Perform Qualitative Risk Analysis process. [Planning]

PMI®, *PMBOK® Guide*, 2013, 331
PMI® *PMP Examination Content Outline,* 2015, Planning, 6, Task 10

11. c. Schedule network diagram and duration estimates

When determining the likelihood of meeting the project's schedule end date through Monte Carlo, the schedule network diagram and duration estimate are used as inputs to the simulation program. Cost risk, on the other hand, uses cost estimates. [Planning]

PMI®, *PMBOK® Guide*, 2013, 340
PMI® *PMP Examination Content Outline,* 2015, Planning, 6, Task 10

12. a. Use a beta or triangular probability distribution

Interviews often are used to help quantify the probability and consequences of risks on project objectives. The type of information collected during the interview depends on the type of probability distribution that is used. As an example, information would be gathered on the optimistic (i.e., low) pessimistic (i.e., high) and most likely scenarios. See Figure 11-13 in the *PMBOK® Guide*, 2013, for an example. A beta or triangular distribution is used widely, as shown in Figure 11-14 in the *PMBOK® Guide*, 2013. Other distributions are uniform, normal, and lognormal. Both interviews and probability distributions are examples of data gathering and representation tools and techniques used in Perform Quantitative Risk Analysis. [Planning]

PMI®, *PMBOK® Guide*, 2013, 336–337
PMI® *PMP Examination Content Outline,* 2015, Planning, 6, Task 10

13. c. Accepts the consequences of the risk event should it occur

    Accepting the consequences of the risk event is categorized as risk acceptance. With this risk response approach, the project team takes no action to reduce the probability of the risk's occurring. Avoidance is a strategy where the team acts to eliminate the threat or protect the project from its impact. It may require changes to the project management plan to eliminate the threat completely. The project manager may isolate the project's objectives from the impact of the risk or change the objective in jeopardy. Examples include extending the schedule, changing the strategy, and reducing scope. The project may even need to be terminated. If these risks occur early in the project, they may be able to be avoided by clarifying objectives, obtaining information, improving communications, or acquiring expertise. [Planning]

    PMI®, *PMBOK® Guide*, 2013, 344–345
    PMI® *PMP Examination Content Outline,* 2015, Planning, 6, Task 10

14. d. 56 percent

    The likelihood is determined by multiplying the probability of event 1 by the probability of event 2. While probability investigates the likelihood that a specific risk will occur, impact investigates the possible effect on the project's objectives such as schedule, quality, or performance. Probability and impact are assessed for each risk in Perform Qualitative Risk Analysis. Risk probabilities and impact are rated according to objectives in the risk management plan. Refer to Figure 11-10 in the *PMBOK® Guide*, 2013 for an example of a probability and impact matrix. [Planning]

    PMI®, *PMBOK® Guide*, 2013, 331–332
    Wideman, M.R., *Project and Program Risk Management: A Guide to Managing Project Risks and Opportunities*, 1992, IV-7
    PMI® *PMP Examination Content Outline,* 2015, Planning, 6, Task 10

15. a. Identifies project assumptions

    Project assumptions, which should be enumerated in the project scope statement, are areas of uncertainty, and as such are potential causes of project risk. Uncertainty of assumptions is evaluated as potential causes of project risk. The scope statement, the WBS, and the WBS Dictionary are part of the scope baseline, an input to Identify Risks. [Planning]

    PMI®, *PMBOK® Guide*, 2013, 322
    PMI® *PMP Examination Content Outline,* 2015, Planning, 6, Task 10

16. c. Watch list

    Even low-priority risks must be monitored. A watch list is used to ensure such risks are tracked for continued monitoring in case their status changes. They are still part of the risk register and are included in updates to project documents, an output of the Plan Risk Response process. [Planning]

    PMI®, *PMBOK® Guide*, 2013, 347
    PMI® *PMP Examination Content Outline,* 2015, Planning, 6, Task 10

17. d. Any risks that cannot be known before they occur

    There is a category of risks that is sometimes called unknown-unknowns, meaning that the risk is not knowable and, therefore, the probability of the risk is also not knowable. For example, your lead technical advisor becoming seriously ill, your offices being ransacked by persons engaged in industrial espionage, or one of your subcontractors wins the lottery and runs off to the Cayman Islands are all examples of risks that are not known before they occur. However, these risks are unknown and cannot be managed proactively and may be assigned a management reserve. [Monitoring and Controlling]

    PMI®, *PMBOK® Guide*, 2013, 310
    PMI® *PMP Examination Content Outline,* 2015, Monitoring and Controlling, 9, Task 4

18. b. Sensitivity analysis

    Sensitivity analysis, as a quantitative risk analysis and modeling technique, helps to determine the risks that have the most potential impact on the project. It examines the extent to which the variations in the project's objectives correlate with variations in different circumstances. It also can be used to examine the extent to which the uncertainty of each project element affects the objective being studied when all other uncertain elements are held at their baseline values. An example is the tornado diagram; please refer to Figure 11-15 in the *PMBOK® Guide*, 2013 for an example. [Planning]

    PMI®, *PMBOK® Guide*, 2013, 338
    Wideman, M.R., *Project and Program Risk Management: A Guide to Managing Project Risks and Opportunities* 1992, C-1 and C-2
    PMI® *PMP Examination Content Outline,* 2015, Planning, 6, Task 10

19. d. $900,000

    EMV = ($2M × 60%) + (–$1.5M × 20%) =
    ($1.2M) + (–$300,000) = $900,000
    As another example of a quantitative risk analysis and modeling technique, EMV is a statistical concept that calculates the average outcome when the future includes scenarios that may or may not happen. This is an example of an opportunity with a positive value. Its use requires a risk-neutral assumption, and it is calculated by multiplying the value of each possible outcome by the probability of occurrence and adding the products together. [Planning]

    Frame, J.D., *The New Project Management: Tools for an Age of Rapid Change, Corporate Reengineering, and Other Business Realities,* 2002, 192
    PMI®, *PMBOK® Guide,* 2013, 339
    PMI® *PMP Examination Content Outline,* 2015, Planning, 6, Task 10

20. c. Active acceptance

    Active acceptance means not only accepting the consequences of a risk but also establishing a plan for dealing with the risk, should it occur. The most common active acceptance strategy is to establish a contingency reserve of time, money, or resources to handle the risks. [Planning]

    PMI®, *PMBOK® Guide,* 2013, 345
    PMI® *PMP Examination Content Outline,* 2015, Planning, 6, Task 10

21. d. Forecast potential deviation of the project at completion from cost and schedule targets

    Earned value is used for monitoring overall project performance against a baseline plan. Outcomes from the analysis may forecast potential deviation of the project at completion from its cost and schedule targets. Deviations from the baseline plan also may indicate the potential impact of threats or opportunities. It is a part of variance analysis, a tool and technique in Control Risks. [Monitoring and Controlling]

    PMI®, *PMBOK® Guide,* 2013, 352
    PMI® *PMP Examination Content Outline,* 2015, Monitoring and Controlling, 9, Task 4

22. b. Assign a relative value to the impact on project objectives if the risk in question occurs

    You can develop relative or numeric, well-defined scales using agreed-upon definitions by the stakeholders. When using a numeric scale, each level of impact has a specific number assigned to it. Each risk then is rated on its probability and impact on an objective if it does occur. Evaluation of each risk's importance and priority is typically done by using a look-up table or a probability and impact matrix, as shown in Figure 11-10 in the *PMBOK® Guide*, 2013. This matrix specifies conditions of probability and impact that lead to ratings of risk as low, moderate, or high priority. Descriptive terms or numeric value can be used. [Planning]

    PMI®, *PMBOK® Guide*, 2013, 331–332
    PMI® *PMP Examination Content Outline,* 2015, Planning, 6, Task 10

23. b. Product of the probability and impact of the risk

    Figure 11-10 in the *PMBOK® Guide*, 2013 shows an example of how an organization can rate a risk for each separate objective. This figure also shows how an organization can develop ways to determine the overall rating for each risk. Threats and opportunities can be shown in the same matrix. The risk score provides a convenient way to compare risks because comparing impacts or probabilities alone is meaningless. It helps guide risk responses. As shown in Figure 11-10, risks with a negative impact on objectives and are in the high-risk or dark gray area of the matrix may require priority action and an aggressive response strategy. Those in the medium gray area or low-risk zone may not require action and may be included in the risk register to be on the watch list or may require some contingency reserve to be added. [Planning]

    PMI®, *PMBOK® Guide*, 2013, 331
    PMI® *PMP Examination Content Outline,* 2015, Planning, 6, Task 10

24. c. Performing the contingency plan

    Corrective action in Control Risks is the process of making changes to bring expected performance in line with the project management plan. These actions consist of contingency plans or workarounds. To use them a change request sometimes is needed and is submitted to the Perform Integrated Change Control process. [Monitoring and Controlling]

    PMI®, *PMBOK® Guide*, 2013, 353
    PMI® *PMP Examination Content Outline,* 2015, Monitoring and Controlling, 9, Task 4

25. b. Forces consideration of the probability of each outcome

    As a graphical way to bring together information, decision-tree analysis quantifies the likelihood of failure or opportunities and places a value on each decision. It is a method of expected monetary value analysis, a quantitative risk analysis and modeling technique. [Planning]

    PMI®, *PMBOK® Guide*, 2013, 339
    Wideman, M.R., *Project and Program Risk Management: A Guide to Managing Project Risks and Opportunities*, 1992, C-2 and C-3
    PMI® *PMP Examination Content Outline,* 2015, Planning, 6, Task 10

26. b. Potential risk response

    The risk register is prepared first in the Identify Risks process as an output. It is a document in which the results of risk analysis and risk response planning are recorded and as they are conducted it contains the outcomes of the other risk management processes. The identified risks are described is as much detail as is reasonable. They may use risk statements in a format such as EVENT may occur causing IMPACT, of if a CAUSE exists, EVENT may led to EFFECT. Potential risk responses also may be identified during the Identify Risks process, and if this is the case they are used as inputs to the Plan Risk Responses process. [Planning]

    PMI®, *PMBOK® Guide*, 2013, 327
    PMI® *PMP Examination Content Outline,* 2015, Planning, 6, Task 10

27. b. Checklists

    Checklists are a tool and a technique of the Identify Risks process and include risks encountered on similar, previous projects based on historical information and knowledge. The lowest level of the RBS can be a checklist. However, while it is easy to prepare a checklist and simple to use, it is impossible to make sure it covers all items relative to the project, and some items on it may not apply to the project. It should be reviewed as the project is closed to incorporate any lessons learned for future projects. [Planning]

    PMI®, *PMBOK® Guide*, 2013, 325
    PMI® *PMP Examination Content Outline,* 2015, Planning, 6, Task 10

28. d. Reducing the probability and/or impact of an adverse risk event to an acceptable threshold

It is often more effective to take early action to reduce probability and/or impact of a risk occurring on a project than attempting to repair the damage after the risk has occurred. Examples include adopting less complex processes, conducting more tests, and choosing a more stable supplier. Another example is using a prototype or targeting links that determine severity when it is not possible to reduce the probability of mitigation, such as by designing redundancy in a system to reduce failure from an original component. [Planning]

PMI®, *PMBOK® Guide*, 2013, 345
PMI® *PMP Examination Content Outline*, 2015, Planning, 6, Task 10

29. a. During the concept phase

Risks are highest at the beginning of a project because the project faces an uncertain future, and the cost of changes is lowest at this time because investments in human and material resources are minimal. Risk and uncertainty decrease during the project as decisions are made and deliverables are accepted. Refer to Figure 2-9 in the *PMBOK® Guide*, 2013, as an example. [Planning]

Frame, J.D., *the New Project Management: Tools for an Age of Rapid Change, Corporate Reengineering, and Other Business Realities,* 2002, 80
PMI®, *PMBOK® Guide*, 2013, 40
Wideman, M.R., *Project and Program Risk Management: A Guide to Managing Project Risks and Opportunities*, 1992, II-1–II-5
PMI® *PMP Examination Content Outline*, 2015, Planning, 6, Task 10

30. c. Schedule management plan and cost management plan

The cost and schedule of a project are two areas significantly affected by risk occurrences. Information on these two areas, because of their quantitative nature, provides excellent input to the Perform Quantitative Risk Analysis process to help determine overall impact and to provide guidelines on establishing and managing risk reserves. [Planning]

PMI®, *PMBOK® Guide*, 2013, 335
PMI® *PMP Examination Content Outline*, 2015, Planning, 6, Task 10

31. c. Define a high-level plan

    Meetings are a tool and technique in the Plan Risk Management process. Typical attendees are the project manager, selected team members, anyone in the organization with authority to manage the risk management planning and execution activities (e.g., a member of a PMO with expertise in this area or a member of the core team with this expertise), and others as needed. In these meetings, high-level plans are defined for conducting risk management activities. They also include cost elements for risk and schedule activities, risk contingency reserve approaches, risk management responsibilities, and templates if available for risk categories and terms. The outputs of these meetings are summarized in the risk management plan. [Planning]

    PMI®, *PMBOK® Guide*, 2013, 316
    PMI® *PMP Examination Content Outline,* 2015, Planning, 6, Task 10

32. b. It is an input to Identify Risks.

    Much of the output from planning in other knowledge areas, such as activity cost and duration estimates, may entail risk and is reviewed during the Identify Risks process. This process requires an understanding of the schedule, cost, and quality management plans found in the project management plan. Estimates that are aggressive or developed with a limited amount of information are even more likely to entail risk and, therefore, must also be an input to the Identify Risks process. By reviewing activity cost estimates, risks can be identified as they provide a quantitative assessment of the likely cost to complete schedule activities, and often are expressed as a range of results showing the degree of risk. This review may result in predictions that indicate whether the estimate is sufficient or insufficient to complete the project; an insufficient estimate could pose a risk to the project. By reviewing activity duration estimates, risks may be identified to the time allowance for the activities in the project as a whole. The range of estimates may indicate relevant degrees of risk. [Planning]

    PMI®, *PMBOK® Guide*, 2013, 322
    PMI® *PMP Examination Content Outline,* 2015, Planning, 6, Task 10

33. c. Risk reassessments are regularly scheduled; risk audits are performed as defined in the project's risk management plan.

Both are tools and techniques in the Control Risks process. Risk reassessment is an ongoing activity by the project team. Risks should be discussed at every status meeting. They are regularly scheduled, and the amount and detail of repetition depends on how well the project is progressing relative to its objectives. Risk audits are performed during the project life cycle to examine and document the effectiveness of risk responses. They are conducted at appropriate frequencies as defined in the risk management plan. They may be included during routine project review meetings or at separate risk audit meetings. [Monitoring and Controlling]

PMI®, *PMBOK® Guide*, 2013, 351
PMI® *PMP Examination Content Outline,* 2015, Monitoring and Controlling, 9, Task 4

34. a. Data quality assessment

Perform Qualitative Risk Analysis requires accurate and unbiased data. The use of low-quality data may result in a qualitative risk analysis that is of little use to the project manager regarding understanding of the risk, data available about the risk, data quality, and data reliability and integrity. Often, the collection of risk information is difficult and consumes more time and resources than planned. Refer to Figure 11-10 in the *PMBOK® Guide*, 2013 for an example to see the values used. In the figure, the numbers in the scale are established when the risk attitude of the organization is defined. [Planning]

PMI®, *PMBOK® Guide*, 2013, 331–332
PMI® *PMP Examination Content Outline,* 2015, Planning, 6, Task 10

35. b. Exploit

Although it might have a negative connotation, exploitation is a strategy used for risks with positive impacts where the organization wants to ensure that the opportunity is realized. This strategy seeks to eliminate uncertainty with a particular upside risk by ensuring the opportunity happens. Other examples, in addition to that in the question, are using new technology or upgrading technology to reduce the cost and duration required to realize the project's objectives. [Planning]

PMI®, *PMBOK® Guide*, 2013, 345
PMI® *PMP Examination Content Outline,* 2015, Planning, 6, Task 10

36. d. Ensuring that each risk identified and deemed critical has a computed expected value

    It is not feasible or necessary to quantify every risk. Therefore, a risk audit should never have as an objective to ensure that each project risk has a computed expected value. However, the format for the audit and its objectives should be defined clearly before the audit is conducted. [Monitoring and Controlling]

    PMI®, *PMBOK® Guide*, 2013, 351
    PMI® *PMP Examination Content Outline,* 2015, Monitoring and Controlling, 9, Task 4

37. a. Defining the steps to be taken if an identified risk event should occur

    Contingent response strategies are a tool and technique in Plan Risk Responses. For some risks it is appropriate for the project team to make a response plan that will be executed only under certain predefined conditions if it is believed that there will be sufficient warning to implement the plan. Events that trigger the contingency response should be defined and tracked. Risk responses identified through this technique may be called contingency plans or fallback plans and include identified triggering events that set the plans into effect. [Planning]

    PMI®, *PMBOK® Guide*, 2013, 346
    PMI® *PMP Examination Content Outline,* 2015, Planning, 6, Task 10

38. b. A risk taken to achieve a reward

    Project risk has its origin in the uncertainty that is present in all projects. Organizations and stakeholders are willing to accept varying degrees of risk, and risks that are threats to the project may be accepted if the risks are within tolerances and are in balance with the rewards to be gained. This example of adopting a fast-track schedule is a risk taken to achieve the reward created by the earlier completion date. [Planning]

    PMI®, *PMBOK® Guide*, 2013, 345
    PMI® *PMP Examination Content Outline,* 2015, Planning, 6, Task 10

39. a. Use a probability and impact matrix

The probability and impact matrix can be used to classify risks according to their level of impact and to prioritize them for future quantitative analyses and responses based on their rating. Typically these risk rating rules are specified by the organization in advance of the project. The matrix specifies combinations of probability and impact that lead to rating the risks as low, moderate, or high priority. See Figure 11-10 for an example in the *PMBOK® Guide*, 2013. [Planning]

PMI®, *PMBOK® Guide*, 2013, 331–332
PMI® *PMP Examination Content Outline,* 2015, Planning, 6, Task 10

40. d. Avoidance

Risk avoidance usually involves changing the project management plan to eliminate the threat entirely. The project manager may isolate the project's objectives from the risk's impact or change the objective in jeopardy. [Planning]

PMI®, *PMBOK® Guide*, 2013, 344
PMI® *PMP Examination Content Outline,* 2015, Planning, 6, Task 10

Project Procurement Management

Study Hints

The Project Procurement Management questions on the PMP® certification exam tend to be more process oriented than legally focused. You do not need to know any country's specific legal code; however, some non-U.S. exam takers complain that the nature of many of the questions requires an understanding of U.S. contract law. Although an occasional question relating to the U.S. system may appear on the exam, such questions do not seem to be problematic for most exam takers. A firm understanding of the procurement process usually will help you to find the correct answer. Moreover, the questions will be worded such that the project manager or project team is the "buyer."

The exam requires you to know the basic differences between the three broad categories of contracts (fixed-price, cost-reimbursement, and time-and-materials) and the risks inherent in specific contract types for both the buyer and the seller. Several questions will also test your knowledge of the various types of contracts within each category (for example, firm-fixed-price versus fixed-price-incentive-fee contracts). A question or two may also be included on international contracting, such as the timing of foreign currency exchange and duty on goods delivered to a foreign country.

PMI® views Project Procurement Management as a four-step process comprising Plan Procurement Management, Conduct Procurements, Control Procurements, and Close Procurements. See *PMBOK® Guide* Figure 12-1 for an overview of this structure. Know this chart thoroughly.

Following is a list of the major Project Procurement Management topics. Use it to help focus your study efforts on the areas most likely to appear on the exam.

Major Topics

Project procurement management overview
Plan Procurement Management

- Make-or-buy analysis
- Procurement management plan
- Procurement statement of work
- Contract categories and risks
  - Fixed-price
  - Cost-reimbursement
  - Time-and-materials
  - Contract types and risks
  - Cost-plus-fixed-fee
  - Cost-plus-incentive-fee
  - Cost-plus-award-fee
  - Fixed-price-incentive-fee
  - Fixed price with economic price adjustments
- Contract incentives
- Contract origination
- Source selection criteria
- Procurement documents
- Make-or-buy decisions
- Change requests

Conduct Procurements

- Proposals
- Qualified seller list
- Evaluating prospective sellers
  - Contract negotiation
  - Weighting system
  - Screening system
  - Independent estimates
  - Proposal evaluation techniques
- Bidder conferences
- Expert judgment
- Analytical techniques
- Contract negotiation stages and tactics
  - Five stages
  - Negotiation tactics
  - Selected sellers
  - Agreements
- Contracts

Control Procurements

- Procurement documents
- Standard clauses
- Elements of a legally enforceable contract
- Changes and change control
- Undefined work
- Procurement performance reviews
- Claims administration
- Records management system

Close Procurements

- Contract closure procedure
- Procurement audit
- Procurement negotiations
- Closed procurements

Practice Questions

INSTRUCTIONS: Note the most suitable answer for each multiple-choice question in the appropriate space on the answer sheet.

1. Assume you are working on a complex project. In fact, it is the most complex project in your organization, so the CEO is interested in it and asks questions about its progress in staff meetings. On this project, you are outsourcing a lot of this work. This means you should—
   a. Award firm-fixed projects
   b. Hold weekly meetings with each contractor to assess progress personally
   c. Manage these multiple contracts in sequence
   d. Have a full-time contract specialist and set up a projectized structure

2. Which term describes those costs in a contract that are associated with two or more projects but are not traceable to either of them individually?
   a. Variable
   b. Direct
   c. Indirect
   d. Semivariable

3. Contract type selection is dependent on the degree of risk or uncertainty facing the project manager. From the perspective of the buyer, the preferred contract type in a low-risk situation is—
   a. Firm-fixed-price
   b. Fixed-price-incentive
   c. Cost-plus-fixed fee
   d. Cost-plus-a-percentage-of-cost

4. The buyer has negotiated a cost-plus-incentive fee contract with the seller. The contract has a target cost of $300,000, a target fee of $40,000, a share ratio of 80/20, a maximum fee of $60,000, and a minimum fee of $10,000. If the seller has actual costs of $380,000, how much fee will the buyer pay?
   a. $104,000
   b. $56,000
   c. $30,000
   d. $24,000

5. Assume you are managing a project and have decided to outsource about 40% of the work. You expect to award three contracts for certain work packages in the WBS. The benefit of this Conduct Procurements process is that it—

   a. Determines the type of contract to award
   b. Aligns internal and external stakeholder expectations through established agreements
   c. Enables a detailed review of the make-or-buy decisions
   d. Avoids the need for additional change requests

6. In some cases, contract termination refers to—

   a. Contract closeout by mutual agreement
   b. Contract closeout by delivery of goods or services
   c. Contract closeout by successful performance
   d. Certification of receipt of final payment

7. Significant differences between the seller's price and your independent estimate may indicate all the following EXCEPT the—

   a. SOW was not adequate
   b. Seller misunderstood the SOW
   c. Sellers failed to respond
   d. Project team chose the wrong contract type

8. You are a contractor for a state agency. Your company recently completed a water resource management project for the state and received payment on its final invoice today. A procurement audit has been conducted. Formal notification that the contract has been closed should be provided to your company by the—

   a. State's project manager
   b. Person responsible for procurement administration
   c. Project control officer
   d. Project sponsor or owner

9. Which term describes contract costs that are traceable to or caused by a specific project work effort?

   a. Variable
   b. Fixed
   c. Indirect
   d. Direct

10. You are awarding a contract to help develop a new feature in the eighth generation phone. You feel this feature will enable your company to outpace the competition. However, you know some potential sellers have worked for competitors in the past. As you select source selection criteria, you should include—

    a. Compensatory damages
    b. Overall life-cycle costs
    c. Intellectual property rights
    d. Warranty

11. Which term is NOT a common name for a procurement document that solicits an offer from prospective sellers?

    a. Seller initial response
    b. Request for information
    c. Request for quotation
    d. Invitation for negotiation

12. It is important to use a records management system on any project. In the procurement management process example of items to archive are—

    a. Contract completion statement
    b. Correspondence
    c. Outstanding claims
    d. Settled subcontracts

13. Recent data indicate that more than 10,000 airline passengers are injured each year from baggage that falls from overhead bins. You performed a make-or-buy analysis and decided to outsource an improved bin design and manufacture it. The project team needs to develop a list of qualified sources. As a general rule, which method would the project team find especially helpful?

    a. Advertising
    b. Internet
    c. Trade catalogs
    d. Relevant local associations

14. As you prepare to close out contracts on your project, you should review all the following types of documentation EXCEPT the—

    a. Contract document for the contract being closed
    b. Procurement audit report
    c. Payment records
    d. Contract changes

15. You are working on a new project in your organization. You need to decide how best to staff the project and handle all its resource requirements. Your first step should be to—

    a. Conduct a make-or-buy analysis
    b. Conduct a market survey
    c. Solicit proposals from sellers using an RFP to determine whether you should outsource the project
    d. Review your procurement department's qualified-seller lists and send an RFP to selected sellers

16. Your company decided to award a contract for project management services on a pharmaceutical research project. Because your company is new to project management and does not understand the full scope of services that may be needed under the contract, it is most appropriate to award a—

    a. Firm-fixed-price contract
    b. Fixed-price-incentive contract
    c. Cost-plus-a-percentage-of-cost contract
    d. Time-and-materials contract

17. Requirements for formal contract acceptance and closure usually are defined in the—

    a. Proposal
    b. Statement of work
    c. Contract terms and conditions
    d. Procurement audit report

18. You plan to award a contract to provide project management training for your company. You decide it is important that any prospective contractor have an association with a major university that awards master's certificates in project management. This is an example of—

    a. Setting up an independent evaluation
    b. Preparing requirements for your statement of work
    c. Establishing a weighting system
    d. Establishing source selection criteria

19. The project team has the responsibility to ensure a procurement agreement—

    a. Incorporates other items the seller specifies
    b. Focuses on all selected sellers deemed to be in the competitive range
    c. Adheres to the negotiation process
    d. Adheres to the organization's procurement policies

20. During the course of working with a seller that no one had worked with previously and is working on site with your team, inspections and audits can be conducted as you specified in the contract. You decided to hire an outside person, considered to be an expert in project management to conduct an audit. In doing so, it is important to note that—

    a. The audit can include buyer management personnel
    b. The audit should be completed quickly so you can process change requests as needed for recommendations
    c. The auditor focuses on completed work rather than work in progress
    d. The auditor practices a policy of 'no surprises' with you and your team with transparent communications

21. You are responsible for ensuring that your seller's performance meets contractual requirements. For effective contract control, you should—

    a. Hold a bidders' conference
    b. Establish the appropriate contract type
    c. Implement the contract change control system
    d. Develop a statement of work

22. The primary benefit of contract control procurements is to ensure that—

    a. Buyers conduct performance reviews
    b. Payment is made in a timely fashion
    c. Disagreements are handled quickly and to everyone's satisfaction
    d. Both parties meet contractual obligations and protect their legal rights

23. An often-overlooked output to the Conduct Procurements process is—

    a. Analytical techniques
    b. Resource calendars
    c. Work performance information
    d. Work performance data

24. Assume you have completed your procurement management plan and your make-or-buy analysis and now are updating project documents as a result of this work. You need to update the—

    a. Stakeholder register
    b. Requirements traceability matrix
    c. Negotiations process
    d. Management systems for contractual relationships

25. You have decided to award a contract to a seller that has provided quality services to your company frequently in the past. Your current project, although somewhat different from previous projects, is similar to other work the seller has performed. In this situation, to minimize your risk you should award what type of contract?

    a. Fixed price with economic price adjustment
    b. Fixed-price-incentive (firm target)
    c. Firm-fixed-price
    d. Cost-plus-award-fee

26. As project manager, you need a relatively fast and informal method addressing disagreements with contractors. One such method is to submit the issue in question to an impartial third party for resolution. This process is known as—

    a. Alternative dispute resolution
    b. Problem processing
    c. Steering resolution
    d. Mediation litigation

27. Work performance data often are used In Control Procurements to show—

    a. Whether quality standards are being satisfied
    b. If approved change requests are being implemented
    c. Current or potential problems
    d. Compliance with contracts

28. In the Plan Procurement Management process, potential sellers are evaluated, especially if the buyer wants to exercise some degree of influence or control over acquisition decisions. This means thought also should be given to—

    a. Cost estimates developed by potential sellers
    b. The scope baseline
    c. How well sellers implement corrective actions as needed
    d. Responsibility to acquire professional licenses

29. A buyer has negotiated a fixed-price-incentive-fee contract with the seller. The contract has a target cost of $200,000, a target profit of $30,000, and a target price of $230,000. The buyer also has negotiated a ceiling price of $270,000 and a share ratio of 70/30. If the seller completes the contract with actual costs of $170,000, how much profit will the buyer pay the seller?

    a. $21,000
    b. $35,000
    c. $39,000
    d. $51,000

30. Requirements for acceptance criteria are defined in the—

    a. Contract
    b. Procurement management plan
    c. Overall project management plan
    d. Specifications

31. Payment systems are a tool and technique in Control Procurements. They are typically processed by—

    a. The contracting officer's technical representative
    b. The contracts office
    c. The accounts payable system
    d. The contractual terms and conditions

32. Assume before you award a contract, you and your team are striving to identify areas that may have more risk. You decide to concentrate on—

    a. Use of an independent estimator
    b. Past performance
    c. Quality levels
    d. Prior agreements

33. Assume in your project, you need to use a contractor. But you and the seller have a prior agreement in place. This means together you can prepare a procurement statement of work. It is an example of—

    a. Core capabilities of the organization
    b. Organizational process assets
    c. Mutual development of contract clauses
    d. Value delivered by vendors

34. The best approach to resolve the settlement of all outstanding contract changes, claims, and disputes is using—

    a. Litigation
    b. Alternative dispute resolution
    c. Negotiation
    d. Mediation

35. Assume on your project, you have a requested but unresolved change as you provided direction to the seller to take action. You consider this change to be one that is—

    a. A claim
    b. A constructive change to the contract
    c. One that requires the need to negotiate a settlement
    d. An undefinitized contractual action

36. You are working on a complex project designed to combat glaucoma without the need to take eye drops. You are going to outsource part of the work to experts in the glaucoma field and have identified four sellers that have been judged to be in a competitive rage. These sellers and you have—

    a. Negotiated a draft contract
    b. Scheduled presentations to your executive team
    c. Established partnering agreements to reduce the possibility of claims
    d. Determined the best approach is a firm-fixed-price contract

37. Assume you thought you would need to outsource part of your work on your telecom project. You conducted a make-or-buy analysis, and the decision was made not to outsource. This means—

    a. You have no additional work remaining in procurement management
    b. The decision should be documented in a decision log
    c. You should update organizational process assets
    d. Your procurement management plan may define processes to follow internal to the organization

38. You are the project manager with a team to design the next generation of automobiles for your company. After an extensive make-or-buy analysis, you have determined which items will be outsourced. Now you are working on the statements of work for each item to be procured. It is important as you do so to—

    a. Determine constraints and assumptions
    b. Follow the scope baseline
    c. Identify any prequalified sellers
    d. Recognize long-lead items

39. Which of the following types of contracts has the least risk to the seller?

    a. Firm-fixed-price
    b. Cost-plus-fixed-fee
    c. Cost-plus-award-fee
    d. Fixed-price-incentive fee

40. Assume that your company has a cost-plus-fixed-fee contract. The contract value is $110,000, which consists of $100,000 of estimated costs with a 10-percent fixed fee. Assume that your company completes the work but only incurs $80,000 in actual cost. What is the total cost to the project?

    a. $80,000
    b. $90,000
    c. $100,000
    d. $125,000

Answer Sheet

1.	a	b	c	d	21.	a	b	c	d
2.	a	b	c	d	22.	a	b	c	d
3.	a	b	c	d	23.	a	b	c	d
4.	a	b	c	d	24.	a	b	c	d
5.	a	b	c	d	25.	a	b	c	d
6.	a	b	c	d	26.	a	b	c	d
7.	a	b	c	d	27.	a	b	c	d
8.	a	b	c	d	28.	a	b	c	d
9.	a	b	c	d	29.	a	b	c	d
10.	a	b	c	d	30.	a	b	c	d
11.	a	b	c	d	31.	a	b	c	d
12.	a	b	c	d	32.	a	b	c	d
13.	a	b	c	d	33.	a	b	c	d
14.	a	b	c	d	34.	a	b	c	d
15.	a	b	c	d	35.	a	b	c	d
16.	a	b	c	d	36.	a	b	c	d
17.	a	b	c	d	37.	a	b	c	d
18.	a	b	c	d	38.	a	b	c	d
19.	a	b	c	d	39.	a	b	c	d
20.	a	b	c	d	40.	a	b	c	d

Answer Key

1. c. Manage these multiple contracts in sequence

   Complex projects often involve contractors and subcontractors. While they may be managed in sequence, they also maybe managed simultaneously. In such cases, each contract life cycle many end during any phase of the project life cycle. Contracts also can be awarded in any phase except in the Close Project phase. [Monitoring and Controlling]

   PMI®, *PMBOK® Guide*, 2013, 357
   PMI® *PMP Examination Content Outline*, 2015, Monitoring and Controlling, 9, Task 7

2. c. Indirect

   The nature of an indirect cost is such that it is neither possible nor practical to measure how much of the cost is attributable to a single project. These costs are allocated to the project by the performing organization as a cost of doing business. In Plan Procurement Management, these costs are considered in the make-or-buy analysis such as the direct costs of supporting the purchasing process and the purchased item. Make-or-buy analysis is a tool and technique in Plan Procurement Management. [Planning]

   PMI®, *PMBOK® Guide*, 2013, 202, 365
   PMI® *PMP Examination Content Outline*, 2015, Planning, 6, Task 7

3. a. Firm-fixed-price

   Buyers prefer the firm-fixed-price contract because it places more risk on the seller. Although the seller bears the greatest degree of risk, it also has the maximum potential for profit. Because the seller receives an agreed-upon amount regardless of its costs, it is motivated to decrease costs by efficient production. It is the most commonly used contract as the price for goods is set at the beginning and is not changed unless the scope of work changes. Types of contracts are an input to the Plan Procurement Management process. [Planning]

   Adams, J.R., et al., *Principles of Project Management*, 1997, 229–231
   PMI®, *PMBOK® Guide*, 2013, 363
   PMI® *PMP Examination Content Outline*, 2015, Planning, 6, Task 7

4. d. $24,000

Comparing actual costs with the target cost shows an $80,000 overrun. The overrun is shared 80/20 (with the buyer's share always listed first). In this case 20% of $80,000 is $16,000, the seller's share, which is deducted from the $40,000 target fee. The remaining $24,000 is the fee paid to the seller. In this type of contract the seller is reimbursed for all allowable costs for performing the contract work and receives a predetermined incentive fee based on achieving certain performance objectives set forth in the contract. If the final costs are less than or greater than the original estimated costs, then the buyer and seller share costs based on a predetermined cost sharing formula, such as the 80/20 as discussed. [Planning and Closing]

PMBOK® Guide, 2013, 364
PMI® *PMP Examination Content Outline,* 2015, Planning, 6, Task 7
PMI® *PMP Examination Content Outline,* 2015, Closing, 11, Task 3

5. b. Aligns internal and external stakeholder expectations through established agreements

The Conduct Procurements process obtains seller responses, selects sellers, and awards contracts. During the process, the team receives bids or proposals and will apply previously defined selection criteria to select the sellers qualified to perform the work. [Executing]

PMBOK® Guide, 2013, 371–372
PMI® *PMP Examination Content Outline,* 2015, Executing, 8, Task 2

6. a. Contract closeout by mutual agreement

A contract can end in successful performance, mutual agreement, or breach of contract. Early termination is a special case of procurement closure that can result from a mutual agreement by both parties, from the default of one party, or for the convenience of the buyer. The rights and responsibilities of both parties for early termination are stated in the terminations clause of the contract. [Closing]

PMBOK® Guide, 2013, 387
PMI® *PMP Examination Content Outline,* 2015, Closing, 11, Task 3

7. d. Project team chose the wrong contract type

   The contract type is typically dictated by the procurement SOW and chosen by the contracting officer. Independent estimates are a tool and technique in Conduct Procurements. The procuring organization may choose to prepare its own independent estimate or have an estimate prepared by an outsider who is an expert in this area to serve as a benchmark on proposed responses. [Executing]

   PMI®, *PMBOK® Guide,* 2013, 376
   PMI® *PMP Examination Content Outline,* 2015, Executing, 8, Task 2

8. b. Person responsible for procurement administration

   The person responsible for procurement administration should provide, in writing, formal notification that the contract has been completed. Requirements for formal acceptance and closeout should be defined in the terms and conditions in the contract and are included in the procurement management plan. [Closing]

   PMI®, *PMBOK® Guide,* 2013, 389
   PMI® *PMP Examination Content Outline,* 2015, Closing, 11, Task 3

9. d. Direct

   Direct costs are always identified with the cost objectives of a specific project and include salaries, travel and living expenses, and supplies in direct support of the project. These costs are considered as part of the make-or-buy analysis, a tool and technique used in Plan Procurement Management. [Planning]

   PMI®, *PMBOK® Guide,* 2013, 202, 207, and 365
   PMI® *PMP Examination Content Outline,* 2015, Planning, 6, Task 7

10. c. Intellectual property rights

   Intellectual property rights should be considered as they refer to whether the seller asserts intellectual property rights in the work processes or services they will use or in the products they will produce for the product. You want to make sure the seller cannot then work on something similar for a competitor. Source selection criteria are an output from the Plan Procurement Management process. [Planning]

   PMI®, *PMBOK® Guide,* 2013, 369
   PMI® *PMP Examination Content Outline,* 2015, Planning, 6, Task 7

11. b. Request for information

Procurement documents are used to solicit proposals from prospective sellers. Terms such as bid, tender, and quotation are generally used if the decision will be based on price, while a term such as proposal is used when there are other considerations. Common terms for different types of documents, in addition to those in this question, are request for proposal and tender notice. Procurement documents are an output of the Plan Procurement Management process. A request for information generally is used by the buyer to have potential sellers propose various pieces of information related to a product, service, or result or to a seller capability. [Planning]

PMI®, *PMBOK® Guide,* 2013, 368
PMI® *PMP Examination Content Outline,* 2015, Planning, 6, Task 7

12. b. Correspondence

The records management system is a tool and technique in Close Procurements. Contract documents and correspondence are archived in it as part of this process. [Closing]

PMI®, *PMBOK® Guide,* 2013, 389
PMI® *PMP Examination Content Outline,* 2015, Closing, 11, Task 3

13. a. Advertising

Advertising is a tool and technique in Conduct Procurements. It enables existing lists of sellers to be expanded by using advertising in general circulation such as in newspapers or specialty trade publications. Online resources also can be used. Some government agencies require public advertising for certain types of procurement items, and most require public advertising or online posting of pending government contracts. [Executing]

PMI®, *PMBOK® Guide,* 2013, 376
PMI® *PMP Examination Content Outline,* 2015, Executing, 8, Task 2

14. b. Procurement audit report

In most organizations, a procurement audit is conducted after the contract has been closed. It is a tool and technique in Close Procurements and is a structured review of the procurement process from the Plan Procurement Management through Close Procurements process. Its objective is to identify successes and failures that warrant recognition in the preparation or administration of other procurement contracts on the project or in the organization. The other items in the question are examples of procurement documents to review, an input to the Close Procurements process. [Closing]

PMI®, *PMBOK® Guide,* 2013, 388
PMI® *PMP Examination Content Outline,* 2015, Closing, 11, Task 3

15. a. Conduct a make-or-buy analysis

A make-or-buy analysis is a Plan Procurement Management tool and technique used to determine whether a particular product, service, or result can be produced or performed cost effectively by the performing organization or should be contracted out to another organization. The analysis includes both direct and indirect costs. Available contract types also are considered as part of the buy analysis. [Planning]

PMI®, *PMBOK® Guide,* 2013, 365
PMI® *PMP Examination Content Outline,* 2015, Planning, 6, Task 7

16. d. Time-and-materials contract

A time-and-materials contract is a type of contract that provides for the acquisition of supplies or services on the basis of direct labor hours, at specified fixed hourly rates for wages, overhead, general and administrative expenses, and profit; and materials at cost, including materials-handling costs. They are a hybrid type of contract in that they contain aspects of cost-reimbursable and fixed-price contracts. They are often used to augment existing staff, acquire experts, and obtain outside support when a precise statement of work cannot be quickly prepared. They fit the scenario in this question and can increase in contract value if required. They are part of organizational process assets, an input to Plan Procurement Management. [Planning]

PMI®, *PMBOK® Guide,* 2013, 364
PMI® *PMP Examination Content Outline,* 2015, Planning, 6, Task 7

17. c. Contract terms and conditions

The contract terms and conditions typically describe the procedure the buyer will employ to close the contract. The Close Procurement process therefore supports the Close Project or Phase process as it ensures all contractual agreements are completed or terminated. [Closing]

PMI®, *PMBOK® Guide*, 2013, 387
PMI® *PMP Examination Content Outline,* 2015, Closing, 11, Task 3

18. d. Establishing source selection criteria

The selection criteria are typically included in procurement documents and are then used to rate or score proposals. They can be objective or subjective. In this specific question, they are more than just price as they are also including areas such as understanding of need, technical capability, management approach, technical approach, production capacity and interest, and past performance. Source selection criteria are an input to Plan Procurement Management. [Planning]

PMI®, *PMBOK® Guide*, 2013, 368–369
PMI® *PMP Examination Content Outline,* 2015, Planning, 6, Task 7

19. d. Adheres to the organization's procurement policies

Agreements are an output of the Conduct Procurements process and include terms and conditions plus other items the buyer specifies as to what the seller is to perform or provide. However, it is the responsibility of the project management team to make certain the agreements meet the needs of the project and adhere to organizational procurement policies. [Executing]

PMI®, *PMBOK® Guide*, 2013, 377
PMI® *PMP Examination Content Outline,* 2015, Executing, 8, Task 2

20. a. The audit can include buyer management personnel

Inspections and audits are tools and techniques in the Control Procurements process. They are required by the buyer and supported by the seller as specified in the procurement contract. As the project is executed, they can verify compliance in the seller's work processes and deliverables. Further, if specified in the contract, some audit teams can include buyer personnel, which is appropriate in this question's scenario because the seller's staff is working on site with the project team. [Monitoring and Controlling]

PMI®, *PMBOK® Guide*, 2013, 383
PMI® *PMP Examination Content Outline,* 2015, Monitoring and Controlling, 9, Task 7

21. c. Implement the contract change control system

    The contract change control system is a tool and technique in Control Procurements. It defines the process to modify the procurement. It includes paperwork, tracking systems, dispute resolution procedures, and approval levels necessary to authorize changes and is integrated with the project's integrated change control system. [Monitoring and Controlling]

    PMI®, *PMBOK® Guide,* 2013, 383
    PMI® *PMP Examination Content Outline,* 2015, Monitoring and Controlling, 9, Task 7

22. d. Both parties meet contractual obligations and protect their legal rights

    Contracts are awarded to obtain goods and services in accordance with the buyer's stated requirements. Although there are multiple purposes in the Control Procurements process, ensuring that the seller delivers what is stated in the contract is of paramount importance. This process involves managing procurement relationships, monitoring contract performance, and making changes and corrections to contracts as needed. [Monitoring and Controlling]

    PMI®, *PMBOK® Guide,* 2013, 379
    PMI® *PMP Examination Content Outline,* 2015, Monitoring and Controlling, 9, Task 7

23. b. Resource calendars

    They are an output of Conduct Procurements as the quantity and availability of contracted resources and the dates on which each specific resource or resource group can be active or is idle is documented. [Executing]

    PMI®, *PMBOK® Guide,* 2013, 378
    PMI® *PMP Examination Content Outline,* 2015, Executing, 8, Task 2

24. b. Requirements traceability matrix

    Other documents to update are the requirements documents and the risk register. The procurement management plan can later include certain requirements that may not have been considered at the time a traceability matrix is prepared. It is important in that it helps ensure each requirement adds business value by linking it to the business and project requirements. Each procurement statement of work, for example, contains requirements that the seller must meet; such other requirements as for-performance bonds or insurance contracts and directions to sellers on developing a WBS are part of the procurement management plan. [Planning]

    PMI®, *PMBOK® Guide,* 2013, 118–119, 367, 370
    PMI® *PMP Examination Content Outline,* 2015, Planning, 6, Task 7

25. c. Firm-fixed-price

In a firm-fixed-price contract, the seller receives a fixed sum of money for the work performed regardless of costs. This arrangement places the greatest financial risk on the seller and encourages it to control costs. Any cost increase because of adverse performance is the responsibility of the seller, who is obligated to complete the effort. The buyer should precisely state the products or services to be procured as any changes to the procurement specification can increase the buyer's cost. [Planning]

Adams et al., *Principles of Project Management,* 1997, 229
PMI®, *PMBOK® Guide,* 2013, 363
PMI® *PMP Examination Content Outline,* 2015, Planning, 6, Task 7

26. a. Alternative dispute resolution

Alternative dispute resolution, or dispute resolution, is a relatively informal way to address differences of opinion on contracts. Its purpose is to address such issues without having to seek formal legal redress through the courts. It is a method that can be defined in advance as part of the procurement award in Conduct Procurements. In terms of claims administration, a tool and technique in Control Procurements, it is an approach to resolve a claim if the buyer and seller cannot agree. It also is used in procurement negotiations, a tool and technique in Close Procurements, if an equitable settlement cannot be reached on outstanding issues, claims, or disputes. [Executing, Monitoring and Controlling, and Closing]

PMI®, *PMBOK® Guide,* 2013, 378, 384, 388
PMI® *PMP Examination Content Outline,* 2015, Executing, 8, Task 2
PMI® *PMP Examination Content Outline,* 2015, Monitoring and Controlling, 9, Task 7
PMI® *PMP Examination Content Outline,* 2015, Closing, 11, Task 3

27. a. Whether quality standards are being satisfied

Work performance data are an input to Control Procurements. Additionally these data show costs that have been incurred or committed, as well as identification of the seller invoices that have been paid. Answer b refers to approved change requests, a separate input to this process, while answers c and d refer to work performance information, an output from this process. [Monitoring and Controlling]

PMI®, *PMBOK® Guide,* 2013, 382
PMI® *PMP Examination Content Outline,* 2015, Monitoring and Controlling, 9, Task 7

28. d. Responsibility to acquire professional licenses

Attention must also be given to the responsibility to obtain or hold any required permits. They all may affect the type of contract to award, and they may be required by legislation, regulation, or organizational policy to execute the project. [Planning]

PMI®, *PMBOK® Guide,* 2013, 360
PMI® *PMP Examination Content Outline,* 2015, Planning, 6, Task 7

29. c. $39,000

To calculate the fee that the buyer must pay, actual costs are compared with the target cost. If actual costs are less than the target cost, the seller will earn profit that is additional to the target profit. If actual costs are more than the target cost, the seller will lose profit from the target profit. The amount of profit is determined by the share ratio (with the buyer's share listed first). In this example, the seller is under target cost by $30,000. That amount will be split 70/30. So the buyer keeps $21,000, and the seller receives an additional $9,000 added to the target profit, which is the incentive. Total fee is $39,000. In this type of contract the buyer and seller have flexibility that allows for deviations from performance with financial incentives in place tied to agreed-upon metrics. These incentives tend to be related to cost, schedule, and technical performance. The performance targets are established at the beginning. The price ceiling is set, and all costs above the price are the seller's responsibility. [Planning]

PMI®, *PMBOK® Guide,* 2013, 363
PMI® *PMP Examination Content Outline,* 2015, Planning, 6, Task 7

30. a. Contract

Agreements are an output of Conduct Procurements. They may be called an understanding, a contract, a subcontract, or a purchase order. A contract is a legal agreement subject to remedy in the courts. An agreement has multiple major components including acceptance criteria. [Executing]

PMI®, *PMBOK® Guide,* 2013, 377
PMI® *PMP Examination Content Outline,* 2015, Executing, 8, Task 2

31. c. The accounts payable system

    After certification of the seller's satisfactory work by an authorized person on the project team, payments to the seller are processed by the buyer's accounts payable system. All payments are made according to the terms of the contract. [Monitoring and Controlling]

    PMI®, *PMBOK® Guide,* 2013, 383
    PMI® *PMP Examination Content Outline,* 2015, Monitoring and Controlling, 9, Task 7

32. b. Past performance

    Analytical techniques are a tool and technique in Conduct Procurements. The goal is that procurements involve a way vendors can bring value through their offerings. These analytical techniques help organizations identify the readiness of a vendor to provide the desired end state, determine the needed cost, and avoid cost overruns. By examining past performance teams may identify areas that have more risk and need to be monitored more closely. [Executing]

    PMI®, *PMBOK® Guide,* 2013, 376
    PMI® *PMP Examination Content Outline,* 2015, Executing, 8, Task 2

33. b. Organizational process assets

    They are an input to Control Procurements. This example is one of a prior arrangement. After the buyer and seller work collectively to prepare a statement of work to satisfy the requirements, they negotiate a final contract to award. Other organizational process assets are listings of prospective and previously qualified sellers and information on relevant positive and negative experiences with sellers. [Executing]

    PMI®, *PMBOK® Guide,* 2013, 375
    PMI® *PMP Examination Content Outline,* 2015, Executing, 8, Task 2

34. c. Negotiation

    While there are a variety of ways to settle claims, disputes, and changes, the preferred approach is negotiation. It is a strategy to work toward compromise or to reach an agreement that both parties can accept. Claims administration is a tool and technique in the Control Procurement process. If the parties cannot negotiate among themselves, alternative dispute resolution is the next step, following specified procedures in the contract. [Monitoring and Controlling]

    PMI®, *PMBOK® Guide,* 2013, 384, 517
    PMI® *PMP Examination Content Outline,* 2015, Monitoring and Controlling, 9, Task 7

35. b. A constructive change to the contract

    This situation is a constructive change to the contract. These constructive changes may be disputed by one party and can lead to a claim against the other party. These changes are uniquely identified and documented by project correspondence. Change requests are an output of Control Procurements and are processed for review and approval through the Integrated Change Control process. [Monitoring and Controlling]

    PMI®, *PMBOK® Guide,* 2013, 385
    PMI® *PMP Examination Content Outline,* 2015, Monitoring and Controlling, 9, Task 7

36. a. Negotiated a draft contract

    Selected sellers are an output of Conduct Procurements. These selected sellers are ones in a competitive range based on the proposal or bid evaluation. They have negotiated a draft contract that will be the actual contract after an award is made. Typically approval of high-value, high-risk, complex procurements generally require approval by senior managers prior to award. [Executing]

    PMI®, *PMBOK® Guide,* 2013, 377
    PMI® *PMP Examination Content Outline,* 2015, Executing, 8, Task 2

37. d. Your procurement management plan may define processes to follow internal to the organization

    Make-or-buy decisions are an output to Plan Procurement Management. If the decision is to make the item, then the procurement management plan may define processes and agreements to follow internal to the organization. If the decision is to buy, the next step is to reach agreement with a seller for the product, service, or results. [Planning]

    PMI®, *PMBOK® Guide,* 2013, 370
    PMI® *PMP Examination Content Outline,* 2015, Planning, 6, Task 7

38. b. Follow the scope baseline

    The procurement statement of work is an output of Plan Procurement Management. The statement of work for each procurement is developed from the project scope baseline and defines the portion of the project's scope to be included within the related contract The other possible answers are items that are part of the procurement management plan. [Planning]

    PMI®, *PMBOK® Guide,* 2013, 367
    PMI® *PMP Examination Content Outline,* 2015, Planning, 6, Task 7

39. b. Cost-plus-fixed-fee

On a firm-fixed-price contract, the seller absorbs 100 percent of the risks; while on a cost-type contract, the buyer carries the most risk. Cost-reimbursable contracts provide flexibility to the project to redirect the seller whenever the scope of work cannot be precisely defined at the start and needs to be altered or when high risks may exist in the effort. Cost-plus-fixed-fee contracts have less risk to sellers than cost-plus-award-fee or cost-plus-incentive-fee contracts because the fee is fixed based on costs, so the seller is guaranteed a certain level of profit. The fee is paid only for completed work and does not change unless the project scope changes. [Planning]

PMI®, *PMBOK® Guide*, 2013, 363–364
PMI® *PMP Examination Content Outline,* 2015, Planning, 6, Task 7

40. b. $90,000

In this situation the fixed-fee of $10,000 does not change but now represents a seller profit of 12.5 percent on incurred costs. This means that the total cost to the project is $90,000. The actual fee is determined when the seller completes its work. [Planning]

Fleming, Q.W., *Project Procurement Management Contracting, Subcontracting, Teaming*, 2003, 97
PMI®, *PMBOK® Guide*, 2013, 363–364
PMI® *PMP Examination Content Outline,* 2015, Planning, 6, Task 7

Project Stakeholder Management

Study Hints

Project Stakeholder Management was added as a tenth knowledge area in the *PMBOK® Guide*—Fifth Edition. The four processes cover four of the five processes in the PMP® certification exam. Stakeholder management is expanded because identifying and analyzing stakeholder expectations and their impact on the project and developing management techniques to effectively engage stakeholders in project decisions and execution are critical to project success. The project manager and his or her team must have a continuous dialogue with stakeholders to meet their needs and expectations, address any issues they may have, and foster the level of appropriate stakeholder engagement in project decisions and activities.

With Project Stakeholder Management as a separate knowledge area, the importance of working with stakeholders on a project is emphasized. It involves focusing on managing the expectations of the project's stakeholder groups and engaging them in the project as appropriate. Research in the project management field further has shown that stakeholder engagement is one of the major keys to project success.

Questions in this knowledge area will address the key stakeholders on projects, as well as areas covered in its four process groups. The four processes not only interact with one another but also interact with processes in the other nine knowledge areas. You need to study these processes carefully to become familiar with PMI®'s terminology and perspectives. *PMBOK® Guide* Figure 13-1 provides an overview of the structure of Project Stakeholder Management. Know this chart thoroughly.

Following is a list of the major Project Stakeholder Management topics. Use it to help focus your study efforts on the areas most likely to appear on the exam.

Major Topics

Stakeholder definition
Types of stakeholders on projects
Identify stakeholders

- Stakeholder analysis
- Preparation
- Classification models
- Meetings
- Stakeholder register

Plan stakeholder management

- Project management plan
- Organizational process assets
- Enterprise environmental factors
- Expert judgment
- Analytical techniques
- Stakeholder engagement assessment matrix
- Contents of the stakeholder management plan

Manage stakeholder engagement

- Key activities
- Communications management pan
- Change log
- Communications methods
- Interpersonal skills
- Management skills
- Issue log
- Organizational process assets updates

Control stakeholder engagement

- Work performance data
- Information management systems
- Work performance information
- Change requests

Practice Questions

INSTRUCTIONS: Note the most suitable answer for each multiple-choice question in the appropriate space on the answer sheet.

1. During your project, you will have a number of different types of meetings. Some will be informational, others will be key updates, and some will be for decision-making purposes. While different attendees will attend each meeting, a best practice to follow is to:
   a. Group stakeholders into categories to determine which ones should attend each meeting
   b. Invite those stakeholders who have a high level of interest in your project to attend each meeting
   c. Be sensitive to the fact that stakeholders often have very different objectives
   d. Recognize that roles and responsibilities may overlap but practice a policy of 'no surprises' and inform your stakeholders about any upcoming meetings

2. You are managing a project with team members located at customer sites on three different continents. You have a number of stakeholders on your project, and most of them are located outside of the corporate office. Who should be responsible for stakeholder management?
   a. A specific team member in each of the three locations
   b. You, because you are the project manager
   c. The project sponsor
   d. A core team including you, as the project manager, and three representatives from the three different locations

3. Analyzing stakeholders is a part of the Identify Stakeholders process. Common approaches for analyzing stakeholders in a qualitative manner includes all the following two-axis grids, EXCEPT—
   a. Comparing power and influence
   b. Comparing power and interest
   c. Comparing influence and location
   d. Comparing influence and impact

4. You are responsible for a project in your organization that has multiple internal customers. Because many people in your organization are interested in this project, you decide to prepare a stakeholder management strategy. Before preparing this strategy, you should—

   a. Conduct a stakeholder analysis to assess information needs
   b. Determine a production schedule to show when each stakeholder needs each type of information produced
   c. Determine the potential impact that each stakeholder may generate
   d. Prioritize each stakeholder's level of interest and influence

5. Recognizing the importance of preparing a stakeholder management plan, you met with your team to obtain their buy in and to discuss it. You explained the key benefit of Plan Stakeholder Management is to—

   a. Determine appropriate strategies for a continual focus on identifying stakeholders throughout the life cycle
   b. Provide a clear plan that is actionable to interact with stakeholders to support the project's interests
   c. Develop appropriate management strategies to effectively engage stakeholders
   d. Plan a series of meetings to ensure stakeholders remain interested and to address their concerns

6. Assume you are actively working, along with your team, to manage stakeholder engagement on your project to develop a new drug to prevent any retina problems of any type. You know you must manage their engagement throughout the project life cycle. This means some organizational process assets will need updating including—

   a. Informal and formal project reports
   b. The stakeholder register
   c. The stakeholder management plan
   d. Work performance information

7. Stakeholders often have issues, and you have asked each of your team members to document them. At each team meeting, you and your team discuss them and determine appropriate responses. You have a project issue log, which is—

   a. Part of the project's lessons learned
   b. Added to the stakeholder register to show which stakeholder raised it
   c. An output from the Manage Stakeholder Engagement process
   d. An output from the Control Stakeholder Engagement process

8. As you work on your project to update its software training classes to focus on an agile approach, you have a number of key stakeholders. As many people and their managers are requesting these classes, your CEO has taken a special interest in your project and has asked you to accelerate your schedule to complete it in two months, rather than in your planned six months, and still have quality offerings. This means as you work to monitor overall project stakeholder relationships, you should—
   a. Provide notifications to stakeholders about status regularly
   b. Ask your stakeholders for regular feedback as you work on your project
   c. Provide presentations to each stakeholder group
   d. Determine how changes will be monitored and controlled

9. As you work with your team to prepare your stakeholder management plan, you decided to develop a stakeholder engagement chart. You set it up so you can—
   a. Show the phase of your project of interest to identified stakeholders
   b. Show gaps between current and desired levels of engagement
   c. Determine which stakeholders you and your team felt were critical to project success but did not know about it
   d. Determine when to involve key stakeholders in your project

10. A number of items in the stakeholder management plan are similar to those in the communications management plan. An example is—
    a. Method for updating and refining the plans as the project progresses and develops
    b. Stakeholder communication requirements for the current project phase
    c. Information to be distributed to stakeholders including language, format, content, and level of detail
    d. Time frame and frequency for the distribution of required information to stakeholders

11. Having worked as a project manager for nine years, you know how important it is to identify the critical stakeholders and not to overlook anyone who has a major influence on your project, even if you do not ever plan to meet with or talk with this individual. As you work with your team, you explain the key benefit of the Identify Stakeholder Process is that it—
    a. Identifies the people, groups, or organizations that could impact or influence project decisions
    b. Shows the interdependencies among project stakeholders to enable classification for how best to involve them on your project
    c. Identifies the appropriate focus for each stakeholder or a group of stakeholders
    d. Shows the potential impact each stakeholder has on project success

12. The last step in the stakeholder analysis process is to—
    a. Determine the organizational culture
    b. Assess how stakeholders probably will respond in various situations
    c. Determine stakeholder roles, interests, and expectations
    d. Evaluate the amount of support each stakeholder could generate

13. You realize that on projects, some stakeholders will not be as receptive as others to your project and actually can be negative from the beginning. Assume you have classified your stakeholders on your project designed to ensure students in your city have access to the best educational resources available, whether in class or on line. Your stakeholder management plan is a sensitive document. Therefore, you need to—
    a. Tell your team to never disclose it to anyone outside the team without consulting you first
    b. Involve your team as you develop it but maintain the final copy yourself
    c. Review the validity of its underlying assumptions
    d. Set up an information distribution system and have each team member sign it for concurrence

14. Stakeholder engagement involves a number of critical activities. An example is—
    a. Ensuring goals are met through negotiation and communications
    b. Developing management strategies to engage them during the project's life cycle
    c. Adjusting strategies and plans to engage stakeholders effectively
    d. Identifying the scope and impact of changes to project stakeholders

15. Work performance information is an output of Control Stakeholder Engagement. It includes a number of items, one of which is—
    a. Change requests
    b. Issue log
    c. Documented lessons learned
    d. Status of deliverables

16. Often in working as a project manager, it is easy to overlook key stakeholders. Assume you work for a device manufacturer and are working as the project manager for the next generation valve replacement. Your company has been a leader in this market, which means you have a lot of lessons learned available to you. Your project is scheduled to last four years. As a best practice, you should—
    a. Work actively with your company's Knowledge Management Officer
    b. Consult regularly with your program manager
    c. Work actively with members of your Governance Board
    d. Work actively with members of your company's Portfolio Review Board

17. Assume you are managing the development of a construction project in your city to replace its five bridges so they are state of the art and meet updated safety standards because they originally were constructed 20 years ago. The design work has been completed. You have awarded subcontracts and are set to begin construction. Today your legal department told you to stop work as you had not consulted them, and there was a critical standard you overlooked during the design process. This example shows—
    a. You need to continually work to engage stakeholders on your project
    b. You should use a RACI chart and have one of your team members work with the legal department throughout the project
    c. You should provide the legal department with a copy of your stakeholder management plan and ask for their representative to sign it and offer any comments
    d. You need to continually identify project stakeholders

18. Assume your construction project is for a small city with only 8,500 people. There has been opposition by many residents to it from the beginning, when the city commissioners approved it. The residents recognize they will be severely impacted as the new bridges are implemented, and during the public hearings before the commissioners' decision, they hired an attorney to state they felt the more cost effective approach was to strengthen the bridges so they met today's safety requirements. Residents now know you have been ordered to stop work, and they have requested a meeting with the commission on Tuesday. This means you should—
    a. Develop a mitigation plan to present at this meeting
    b. Work diligently with the legal department to satisfy their concerns and receive a go ahead before Tuesday's meeting
    c. Demonstrate at the meeting the sustainability impacts of the new bridges
    d. Balance the interests of these negative stakeholders and meet with them before Tuesday's meeting

19. The salience model is one way to classify stakeholders. In it—
    a. Stakeholders' power, urgency, and legitimacy are used
    b. Stakeholders' level of authority and concern are used
    c. Stakeholders' active involvement and power are used
    d. Stakeholders' influence and ability to effect changes are used

20. In Plan Stakeholder Management, all organizational assets are used as inputs; however, which of the following are of particular importance?
    a. Organization culture and the political climate
    b. Practices and habits and templates
    c. Lessons learned database and historical information
    d. Organization's knowledge management system and policies and procedures

21. Assume you have identified your stakeholders and are preparing your stakeholder management plan. You are fortunate that your team is a colocated team as you are working on an internal project to reorganize your IT Department so it is focused more on its customers. The project sponsor is the Chief Operating Officer, and the IT Department Director was surprised as she thought all was well. However, you notice when planning meetings are held, the Chief Financial Officer never attends. You believe that IT affects the entire company and therefore all the senior leaders need some type of involvement. You therefore feel the Chief Financial Officer may be—
    a. Resistant
    b. Unaware
    c. Uninterested
    d. Satisfied

22. Assume your stakeholder management plan has been approved. You now are working with your team to promote stakeholder engagement on your project. You explain in a team meeting its benefit is to—
    a. Clarify and resolve identified issues
    b. Meet stakeholder needs and expectations
    c. Obtain their continued commitment to the project
    d. Increase support and minimize resistance

23. The stakeholder register should not be prepared only one time, but it should be updated regularly especially if—
    a. The stakeholder is not an active participant
    b. The stakeholder is not impacted by the project
    c. The stakeholder does not read status updates
    d. The stakeholder leads a corporate reorganization

24. You are a project manager working to foster stakeholder engagement, and you know a combination of interpersonal skills and general management skills is needed. An example of a key interpersonal skill in stakeholder engagement is—

    a. Facilitating consensus
    b. Influencing people
    c. Resolving conflicts
    d. Negotiating agreements

25. Stakeholder engagement must be controlled on a continuous basis for it to be effective. You realize a number of project documents can be useful for you as a project manager. An example is—

    a. Technical performance measures
    b. Change log
    c. Actual costs
    d. Start and finish dates of schedule activities

26. Expert judgment is a best practice as a tool and technique in many project management processes, and the list of possible sources for experts varies by the organization and by its association with others. Once you have identified experts who you feel could be of assistance, you can—

    a. Use a focus group
    b. Review documentation
    c. Hold brainstorming sessions with your team
    d. Conduct interviews

27. Enterprise environmental factors are useful in many projects. In Plan Stakeholder Management an example of one of particular importance is—

    a. Lessons learned
    b. Templates from previous stakeholder management plans from successful projects
    c. Benchmarking
    d. Organizational culture

28. Stakeholder management involves more than improving communications and requires more than just managing a team. This is because it—

    a. Contains detailed plans to ensure stakeholder requirements are satisfied
    b. Describes how to best meet human resource requirements
    c. Creates and maintains relationships between the project team and stakeholders
    d. Focuses on active participation of all stakeholders, even the resistors to the project

29. Assume you are beginning your project to develop a series of residential condominiums in your city and are identifying possible stakeholders. A key organizational process asset you can review is—
    a. Organizational culture
    b. Organizational standards
    c. Lessons learned
    d. Local trends

30. One way to develop an understanding of major project stakeholders to exchange and analyze project information about roles and interests is to—
    a. Conduct interviews
    b. Hold profile analysis meetings
    c. Use questionnaires and surveys
    d. Conduct a stakeholder analysis and analyze the results with a focus group

31. Assume you are managing a project to implement an electronic medical record system in your ophthalmologist's office. You have been working to identify your stakeholders to then make sure everyone is committed to it as some people have been working in this office for more than 20 years and are comfortable with the manual approach. At this point, you have documented assessment information, which includes—
    a. Role in the project
    b. Whether the stakeholder is a supporter, is neutral, or is resistant
    c. Potential influence in the project
    d. Organization position

32. Having prepared stakeholder management plans on previous projects, you know it is positive to review the project management plan because it—
    a. Provides information as to how to plan appropriate ways to engage stakeholders
    b. Contains information useful to ensure the stakeholder management plan is aligned with the organization's culture
    c. Helps to determine the best options to support an adaptive process for stakeholder management
    d. Contains a change management plan and documents how changes will be monitored and controlled

33. Assume you have performed your stakeholder analysis and now are working to enhance it with a stakeholder engagement assessment matrix. Such a matrix shows the stakeholder's current engagement level. These data enable—

    a. The project manager to prepare the stakeholder management plan
    b. The project manager to prepare the stakeholder management strategy
    c. The project manager to prepare the stakeholder inventory
    d. The project team to expand the stakeholder risk register

34. The ability of stakeholders to influence a project is—

    a. Constant throughout the project life cycle as different stakeholders have different levels of interest in the project at different times
    b. Highest during the closing stage since key stakeholder acceptance criteria must be met
    c. Highest during planning as the team is still in the storming stage as various stakeholders' positions are being known and recognized
    d. Highest in the very early stages as the project is being approved and initiated

35. Working on your project to design and construct five new bridges for your City, you are striving to actively manage the stakeholders on your project, especially those who will be inconvenienced by the project and have indicated they do not support it. You decided to review your communications management plan as it—

    a. Contains issue management procedures
    b. Describes the project's life cycle and the processes to be used in each phase
    c. Sets forth an escalation process
    d. Provides guidance as to how to best involve stakeholders in the project

36. A supporting input for controlling stakeholder engagement is—

    a. Budget
    b. Project schedule
    c. Historical information
    d. Number of defects

37. As a result of the Control Stakeholder Engagement process, you realize even though this process is under way until the closing phase that you have identified the root cause of some issues you have faced in controlling stakeholders expectations. You should therefore—

    a. Review them with your Governance Board
    b. Revise and reissue your stakeholder management plan
    c. Prepare a change request
    d. Update the lessons learned documentation

38. Identifying interrelationships and potential overlap between stakeholders is useful to the project manager as he or she works with stakeholders. It should be documented as part of the—

    a. Stakeholder register
    b. Stakeholder management strategy
    c. Stakeholder management plan
    d. Stakeholder engagement assessment matrix

39. A number of organizational process assets are useful as inputs to the Manage Stakeholder Engagement process. Similarly a number of organizational process assets require updates because of this process. An example of one that is an input is—

    a. Project reports
    b. Historical information
    c. Project records
    d. Stakeholder notifications

40. Table reporting, spreadsheet analysis, and presentations are examples of—

    a. Project reports as an input to Manage Stakeholder Engagement
    b. Work performance information as an output of Control Stakeholder Engagement
    c. Tools and techniques used in Control Stakeholder Engagement
    d. Updates from the Plan Stakeholder Management process

Answer Sheet

1.	a	b	c	d	21.	a	b	c	d
2.	a	b	c	d	22.	a	b	c	d
3.	a	b	c	d	23.	a	b	c	d
4.	a	b	c	d	24.	a	b	c	d
5.	a	b	c	d	25.	a	b	c	d
6.	a	b	c	d	26.	a	b	c	d
7.	a	b	c	d	27.	a	b	c	d
8.	a	b	c	d	28.	a	b	c	d
9.	a	b	c	d	29.	a	b	c	d
10.	a	b	c	d	30.	a	b	c	d
11.	a	b	c	d	31.	a	b	c	d
12.	a	b	c	d	32.	a	b	c	d
13.	a	b	c	d	33.	a	b	c	d
14.	a	b	c	d	34.	a	b	c	d
15.	a	b	c	d	35.	a	b	c	d
16.	a	b	c	d	36.	a	b	c	d
17.	a	b	c	d	37.	a	b	c	d
18.	a	b	c	d	38.	a	b	c	d
19.	a	b	c	d	39.	a	b	c	d
20.	a	b	c	d	40.	a	b	c	d

Answer Key

1. c. Be sensitive to the fact that different stakeholders often have very different objectives

   A project stakeholder is an individual, group, or organization that is actively involved in the project or have interests that may be affected, either positively or negatively, as a result of the performance or completion of the project. Stakeholders also may exert influence on the project and its results. Managing stakeholder expectations is difficult since stakeholders often have different or conflicting objectives. [Monitoring and Controlling]

   PMI®, *PMBOK® Guide*, 2013, 30
   PMI® *PMP Examination Content Outline,* 2015, Monitoring and Controlling, 9, Task 1

2. b. You, because you are the project manager

   Stakeholder management refers to any action taken by the project manager or project team to satisfy the needs of and to resolve issues with project stakeholders. The ability of the project manager to correctly identify and manage stakeholders appropriately can mean the difference between project success and failure. [Monitoring and Controlling]

   PMI®, *PMBOK® Guide*, 2013, 391
   PMI® *PMP Examination Content Outline,* 2015, Monitoring and Controlling, 9, Task 1

3. c. Comparing influence and location

   Identifying and analyzing the stakeholders helps to classify them better for developing a strategy to help manage them and their expectations throughout the project. The most common comparison elements are: power, influence, interest, and impact. The location of the person may have an impact on one of the other measures, but it is not easily quantifiable on a low-, medium-, high-type scale. [Initiating]

   PMI®, *PMBOK® Guide*, 2013, 396
   PMI® *PMP Examination Content Outline,* 2015, Initiating, 5, Task 3

4. a. Conduct a stakeholder analysis to assess information needs

   Stakeholder analysis is used to analyze the information needs of the stakeholders and to determine the sources for meeting those needs. It helps to determine whose interests should be taken into account throughout the project. It is a tool and technique in the Identify Stakeholders process. [Initiating]

   PMI®, *PMBOK® Guide*, 2013, 395–396
   PMI® *PMP Examination Content Outline,* 2015, Initiating, 5, Task 3

5. b. Provide a clear plan that is actionable to interact with stakeholders to support the project's interests

   While the Plan Stakeholder Management process develops appropriate management strategies to effectively engage the stakeholders during the project life cycle, the key benefit of this process is to have a plan that is clear and actionable to interact with them to support the project's interests. [Planning]

   PMI®, *PMBOK® Guide*, 2013, 399
   PMI® *PMP Examination Content Outline,* 2015, Planning, 6, Task 13

6. a. Informal and formal project reports

   While a number of different organizational process assets require updates as a result of the Manage Stakeholder Engagement process, project reports is one example. They include the formal and informal project reports that describe project status and include lessons learned, issue logs, project closure reports, and outputs from other knowledge areas. Other organizational process assets that may require updates are stakeholder notifications, project presentations, project records, feedback from stakeholders, and lessons learned documentation. [Executing]

   PMI®, *PMBOK® Guide*, 2013, 409
   PMI® *PMP Examination Content Outline,* 2015, Executing, 8, Task 7

7. c. An output from the Manage Stakeholder Engagement process

   Issues logs are an output of this process, as issues are expected in this process. The log is updated as new issues are identified, and existing issues are resolved. Other outputs are: change requests, updates to the project management plan, updates to project documents, and updates to organizational process assets. [Executing]

   PMI®, *PMBOK® Guide*, 2013, 408–409
   PMI® *PMP Examination Content Outline,* 2015, Executing, 8, Task 7

8. d. Determine how changes will be monitored and controlled

    As you work in managing stakeholder engagement you should review your project management plan. Your CEO has requested a major schedule change; among other things the project management plan is an input to this process as it contains a change management plan that documents how changes will be monitored and controlled. It also describes the project life cycle and the processes for each phase, the work to accomplish project objectives, how human resource requirements will be met, and needs and techniques to communicate among stakeholders. [Monitoring and Controlling]

    PMI®, *PMBOK® Guide*, 2013, 411
    PMI® *PMP Examination Content Outline,* 2015, Monitoring and Controlling, 9, Task 1

9. b. Show gaps between current and desired levels of engagement

    The stakeholder engagement assessment matrix is used as a tool and technique in Plan Stakeholder Management. The purpose of the matrix is to show gaps between current and desired engagement levels to then ensure the plan provides these data. Review Figure 13-7 in the *PMBOK® Guide*, 2013 for an example. [Planning]

    PMI®, *PMBOK® Guide*, 2013, 402–403
    PMI® *PMP Examination Content Outline,* 2015, Planning, 6, Task 13

10. a. Method for updating and refining the plans as the project progresses and develops

    The other items listed have specific stakeholder references that, while similar, are not in the communications management plan. This is because the stakeholder management plan's focus is to have a management strategy to effectively engage stakeholders on the project. In contrast, the emphasis of the communications management plan is to describe how project communications will be planned, structured, monitored, and controlled. [Planning]

    PMI®, *PMBOK® Guide*, 2013, 296, 403
    PMI® *PMP Examination Content Outline,* 2015, Planning, 6, Task 13

11. c. Identifies the appropriate focus for each stakeholder or a group of stakeholders

    The Identify Stakeholder process has a number of purposes. It identifies people, groups, or organizations that could impact or be impacted by a decision, activity, or outcome of the project. It analyzes and documents relevant information concerning their interests, involvement, interdependencies, influence and potential impact on project success. Its key benefit is to allow the project manager to identify the appropriate focus for each stakeholder. [Initiating]

    PMI®, *PMBOK® Guide*, 2013, 393
    PMI® *PMP Examination Content Outline,* 2015, Initiating, 5, Task 3

12. b. Assess how stakeholders probably will respond in various situations

    In stakeholder analysis, the last step is to assess how key stakeholders are likely to react or respond to various situations in order to plan how to influence them to enhance their support and mitigate any potential negative impacts. The first step is to identify the relevant stakeholders, and the second step involves analyzing their potential impact or support. **Note:** as you study for the PMP® exam, recognize when steps in a process are discussed, such as in this question, it is important to know the steps as you may see a question such as this one or one about the first step in the process. [Initiating]

    PMI®, *PMBOK® Guide*, 2013, 396
    PMI® *PMP Examination Content Outline,* 2015, Initiating, 5, Task 3

13. c. Review the validity of its underlying assumptions

    Information on resistant stakeholders can be damaging, and consideration is needed regarding distributing the stakeholder management plan and the stakeholder register. The project manager needs to be aware of the sensitive nature of these documents. When preparing and updating them, the best practice is to review the underlying assumptions to ensure continued accuracy and relevancy. [Planning]

    PMI®, *PMBOK® Guide*, 2013, 404
    PMI® *PMP Examination Content Outline,* 2015, Planning, 6, Task 13

14. a. Ensuring goals are met through negotiation and communications

A key activity in Manage Stakeholder Engagement is to manage stakeholder expectations through negotiation and communications, ensuring project goals are achieved. This process focuses on communicating with stakeholders and working with them to meet their needs or expectations, addressing issues as they occur, and fostering appropriate engagement in project activities throughout the project. [Executing]

PMI®, *PMBOK® Guide*, 2013, 404–405
PMI® *PMP Examination Content Outline,* 2015, Executing, 8, Task 7

15. d. Status of deliverables

Work performance information is performance data collected from various controlling processes that are analyzed and integrated based on relationships among areas. The data are transformed into information, which is correlated and contextualized and provides a sound foundation for project decisions. The status of deliverables is an example. Other examples include implementation status for change requests and forecasted estimates to complete. [Monitoring and Controlling]

PMI®, *PMBOK® Guide*, 2013, 413
PMI® *PMP Examination Content Outline,* 2015, Monitoring and Controlling, 9, Task 1

16. c. Work actively with members of your Governance Board

New product development organizations are noted for setting up Governance Boards to oversee projects. Additionally in this situation, it is a long project that is important to the company. Project governance ensures the alignment of the project with stakeholder needs and expectations and is critical to the management of stakeholder expectations and to the achievement of organizational objectives. [Executing]

PMI®, *PMBOK® Guide*, 2013, 30
PMI® *PMP Examination Content Outline,* 2015, Executing, 8, Task 7

17. d. You need to continually identify project stakeholders

    Stakeholder identification is a continual process throughout the project life cycle. The legal department often is overlooked, but it is a significant stakeholder, and in this situation, delays resulted. Significant expenses often are due to legal requirements that must be met before the project can be completed, or the project scope is delivered. As an output of Identify Stakeholders, the stakeholder register should be consulted and updated; stakeholders may change, or new stakeholders will be identified. These updates continue throughout the project. [Initiating]

    PMI®, *PMBOK® Guide*, 2013, 31, 398
    PMI® *PMP Examination Content Outline,* 2015, Initiating, 5, Task 3

18. d. Balance the interests of these negative stakeholders and meet with them before Tuesday's meeting

    Overlooking negative stakeholders' interests can result in an increased likelihood of failures, delays, or other negative consequences to projects. The project manager must control stakeholder engagement, which can be difficult since they often have different or competing objectives. The key benefit of the Control Stakeholder Engagement process is to maintain or increase the effectiveness of stakeholder engagement activities as the project evolves and the project environment changes. Therefore, it is necessary to monitor stakeholder relationships and adjust strategies and plans to engage stakeholders. [Monitoring and Controlling]

    PMI®, *PMBOK® Guide*, 2013, 32, 409
    PMI® *PMP Examination Content Outline,* 2015, Monitoring and Controlling, 9, Task 1

19. a. Stakeholders' power, urgency, and legitimacy are used

    In the salience model, stakeholders are described in classes based on their power or ability to impose their will, urgency or need for immediate action, and legitimacy or their involvement. It is an approach used for stakeholder analysis, along with power/interest, power/influence, and influence/impact. Stakeholder analysis is a tool and technique in Identify Stakeholders. [Initiating]

    PMI®, *PMBOK® Guide*, 2013, 396
    PMI® *PMP Examination Content Outline,* 2015, Initiating, 5, Task 3

20. c. Lessons learned database and historical information

    While it is rare that all organizational process assets are used in any process, these are of particular importance as they provide insight on previous stakeholder management plans and their effectiveness. They can be used to plan stakeholder management activities for the current project. [Planning]

    PMI®, *PMBOK® Guide*, 2013, 401
    PMI® *PMP Examination Content Outline,* 2015, Planning, 6, Task 13

21. a. Resistant

    Since the Chief Financial Officer has financial responsibility for all of the company's work, in preparing a stakeholder engagement strategy, he or she probably is aware of this project, and probably is resistant to change, perhaps feeling resources could be better spent on other initiatives. Analytical techniques are a tool and technique in Plan Stakeholder Management. Other ways to classify stakeholder engagement levels are: unaware, neutral, supportive, or leading. [Planning]

    PMI®, *PMBOK® Guide*, 2013, 402
    PMI® *PMP Examination Content Outline,* 2015, Planning, 6, Task 13

22. d. Increase support and minimize resistance

    The other possible answers are activities in the Manage Stakeholder Engagement process. Its benefit is to allow the project manager to increase support and minimize resistance from stakeholders to significantly increase chances for success. [Executing]

    PMI®, *PMBOK® Guide*, 2013, 404–405
    PMI® *PMP Examination Content Outline,* 2015, Executing, 8, Task 7

23. b. The stakeholder is not impacted by the project

    Project document updates are an output to the Manage Stakeholder Engagement process. These updates involve the stakeholder register. It should be updated as stakeholder information changes, when new stakeholders are identified, or if stakeholders listed in the register are no longer involved in or impacted by the project. [Executing]

    PMI®, *PMBOK® Guide*, 2013, 409
    PMI® *PMP Examination Content Outline,* 2015, Executing, 8, Task 7

24. c. Resolving conflicts

Conflicts are common on projects and between stakeholders. Other interpersonal skills useful in Manage Stakeholder Engagement are building trust, active listening and overcoming resistance to change. The other answers are examples of management sills, another tool and technique in this process. [Executing]

PMI®, *PMBOK® Guide*, 2013, 407–408
PMI® *PMP Examination Content Outline,* 2015, Executing, 8, Task 7

25. b. Change log

Projects involve change, and most everyone tends to resist it. A change log is a project document to review as an input to the Control Stakeholder Engagement process. Other useful documents are the schedule, stakeholder register, issue log, and project communications. The other answers to this question are examples of work performance data, another input to this process. [Monitoring and Controlling]

PMI®, *PMBOK® Guide*, 2013, 412–413
PMI® *PMP Examination Content Outline,* 2015, Monitoring and Controlling, 9, Task 1

26. a. Use a focus group

The key words in the question are "identified experts". Useful ways to obtain information from experts are individual consultations, which include one-on-one meetings and interviews, or through a panel approach, such as focus groups and surveys. Focus groups are an excellent way to obtain insight into attitudes, useful to Control Stakeholder Engagement. [Monitoring and Controlling]

PMI®, *PMBOK® Guide*, 2013, 412
PMI® *PMP Examination Content Outline,* 2015, Monitoring and Controlling, 9, Task 1

27. d. Organizational culture

Enterprise environmental factors are inputs to Plan Stakeholder Management because the management of stakeholders should be adapted to the project environment. This means organizational culture is of particular importance, along with structure and the political climate. These three help determine the best option to support a better and more adaptive approach for managing stakeholders. [Planning]

PMI®, *PMBOK® Guide*, 2013, 401
PMI® *PMP Examination Content Outline,* 2015, Planning, 6, Task 13

28. c. Creates and maintains relationships between the project team and stakeholders

    Plan Stakeholder Management identifies how the project will affect stakeholders, which then allows the project manager to determine how best to involve stakeholders in the project, manage their expectations, and achieve project objectives. It is about creating and maintaining relationships between the project team and stakeholders, with the goal of satisfying their respective needs and requirements within the project's boundaries. Stakeholder management planning is iterative and occurs throughout the project and is reviewed by the project manager. [Planning]

    PMI®, *PMBOK® Guide*, 2013, 400
    PMI® *PMP Examination Content Outline,* 2015, Planning, 6, Task 13

29. c. Lessons learned

    Lessons learned, stakeholder register templates, and stakeholder registers from previous projects are examples of organizational process assets that can influence the Identify Stakeholders process. The other answers are examples of enterprise environmental factors to consider. [Initiating]

    PMI®, *PMBOK® Guide*, 2013, 395
    PMI® *PMP Examination Content Outline,* 2015, Initiating, 5, Task 3

30. b. Hold profile analysis meetings

    A profile analysis meeting is a tool and technique in the Identify Stakeholders process. Its purpose is to develop a deeper understanding of major project stakeholders. The meetings can be used to exchange and analyze information about roles, interests, knowledge, and the overall position of each stakeholder about the project. [Initiating]

    PMI®, *PMBOK® Guide*, 2013, 398
    PMI® *PMP Examination Content Outline,* 2015, Initiating, 5, Task 3

31. c. Potential influence in the project

    The stakeholder register contains assessment information as a key component. The assessment information includes: major requirements, main expectations, potential influence in the project, and the phase in the project life cycle with the most interest. The other answers in this question are examples of items to be included in the stakeholder register but in the category of identification information. [Initiating]

    PMI®, *PMBOK® Guide*, 2013, 398
    PMI® *PMP Examination Content Outline,* 2015, Initiating, 5, Task 3

32. d. Contains a change management plan and documents how changes will be monitored and controlled

Among other key items useful in the project management plan to review while preparing the stakeholder management plan is the change management plan. All projects involve some type of change. Reviewing this plan can help the project manager work with stakeholders who may be resistant to the project to help turn them into ones who are supportive or at least neutral to the resulting changes. Other items to review are the project life cycle and specific processes, methods to describe the work needed to accomplish the project's objectives, ways to best meet human resource management requirements, and the need and techniques for stakeholder communications. [Planning]

PMI®, *PMBOK® Guide*, 2013, 400
PMI® *PMP Examination Content Outline,* 2015, Planning, 6, Task 13

33. a. The project manager to prepare the stakeholder management plan

The stakeholder engagement assessment matrix shows the stakeholders current engagement in the project, and the project manager and team then can use it to note the desired level of engagement. See Figure 13-7 in the *PMBOK® Guide*, 2013 for an example. As a tool and technique in Plan Stakeholder Management, the project manager then uses it to help prepare the stakeholder management plan. [Planning]

PMI®, *PMBOK® Guide*, 2013, 402–403
PMI® *PMP Examination Content Outline,* 2015, Planning, 6, Task 13

34. d. Highest in the very early stages as the project is being approved and initiated

The stakeholders' ability to influence the project is highest during the initial phases and gets progressively lower as the project progresses. Active management of stakeholders' involvement decreases the risk of the project failing to meet its goals and objectives. The project manager is responsible to engage and manage stakeholders and may call upon the sponsor as needed. Active management of stakeholder involvement decreases the risk of the project not meeting its goals and objectives. [Executing]

PMI®, *PMBOK® Guide*, 2013, 406
PMI® *PMP Examination Content Outline,* 2015, Executing, 7, Task 7

35. c. Sets forth an escalation process

    Among other things, an escalation process is helpful especially if there are issues or risks involving communications that the project manager wishes to escalate to determine the most appropriate response or to share the approach he or she plans to follow. Other useful items from the communications management plan include: stakeholder communications requirements, information to be communicated, reasons to distribute information, and the people responsible to receive the information. [Executing]

    PMI®, *PMBOK® Guide*, 2013, 406
    PMI® *PMP Examination Content Outline,* 2015, Executing, 8, Task 7

36. b. Project schedule

    Project documents are an input to Control Stakeholder Engagement. They originate from initiating, planning, executing, or controlling processes and include the project schedule, stakeholder register, issue log, change log, and project communications. [Monitoring and Controlling]

    PMI®, *PMBOK® Guide*, 2013, 411–412
    PMI® *PMP Examination Content Outline,* 2015, Monitoring and Controlling, 9, Task 1

37. d. Update the lessons learned documentation

    This documentation is an example of an organizational process asset to update as it includes the root cause analysis of issues faced, the reasons certain corrective actions were taken, and other types of lessons learned about stakeholder management. These lessons learned are documented and distributed so they are part of the historical database for the project and the performing organization. Other organizational process assets to update are: stakeholder notifications, project reports, project presentations, project records, and stakeholder feedback. [Monitoring and Controlling]

    PMI®, *PMBOK® Guide*, 2013, 415
    PMI® *PMP Examination Content Outline,* 2015, Monitoring and Controlling, 9, Task 1

38. c. Stakeholder management plan

The stakeholder management plan identifies the management strategies required to effectively engage stakeholders. It includes, among other things, the identified interrelationships and potential overlap between stakeholders. This information is invaluable especially if some stakeholders are resistant or negative to the project and also in determining the level of frequency of desired interaction and communications requirements. [Planning]

PMI®, *PMBOK® Guide*, 2013, 403
PMI® *PMP Examination Content Outline,* 2015, Planning, 6, Task 13

39. b. Historical information

Historical information about previous projects, organizational communications requirements, issue management procedures, and change control procedures are examples of organizational process assets that are inputs to the Manage Stakeholder Engagement process. [Executing]

PMI®, *PMBOK® Guide*, 2013, 407
PMI® *PMP Examination Content Outline,* 2015, Executing, 8, Task 7

40. c. Tools and techniques used in Control Stakeholder Engagement

In Control Stakeholder Engagement, they are examples of distribution formats from information management systems, a tool and technique in this process. Such systems provide a structured tool for the project manager to capture, store, and distribute information to stakeholders about project cost, schedule progress, and performance. The project manager can use these systems to consolidate reports from several systems and facilitate report distribution. [Monitoring and Controlling]

PMI®, *PMBOK® Guide*, 2013, 412
PMI® *PMP Examination Content Outline,* 2015, Monitoring and Controlling, 9, Task 1

An Overview of the Domains

The *PMBOK® Guide* is organized according to the ten knowledge areas, but the PMP® exam is based on the domains or process groups. This section provides information about these domains and the examination content outline (ECO) to assist you as you study for the exam. Review Figure A1-2 from the *PMBOK® Guide*, 2013 to see the interactions beween these process groups.

The questions on the exam are distributed as follows by process group following the *PMI® PMP® Examination Content Outline – June 2015*:

- 13% relate to Initiating the Project
- 24% relate to Planning the Project
- 31% relate to Executing the Project
- 25% relate to Monitoring and Controlling the Project
- 7% relate to Closing the Project

Initiating the Project

This domain or process group involves processes to define a new project or phase of an existing project by obtaining authorization to start it. Although it constitutes 13% of the questions on the exam, it involves only two of the 47 processes in the *PMBOK® Guide*. Please see the Figure 11-1 in this chapter.

The key document is the project charter, and you should be familiar with its contents. It is assumed in the *PMBOK® Guide* that the business case assessment, approval, and funding are handled externally to the boundaries of the project. Read sections 1.4.2 on Portfolio Management and 1.4.3 on Projects and Strategic Planning in the *PMBOK® Guide* for additional information.

During the Initiating process, the formal authorization to start a project—or phase if it is a large project—is obtained, the project manager is assigned and is given the authority and responsibility to apply organizational resources to the project, and the key internal and external stakeholders—those who will be involved in, have an influence over, or affected by the project—are identified.

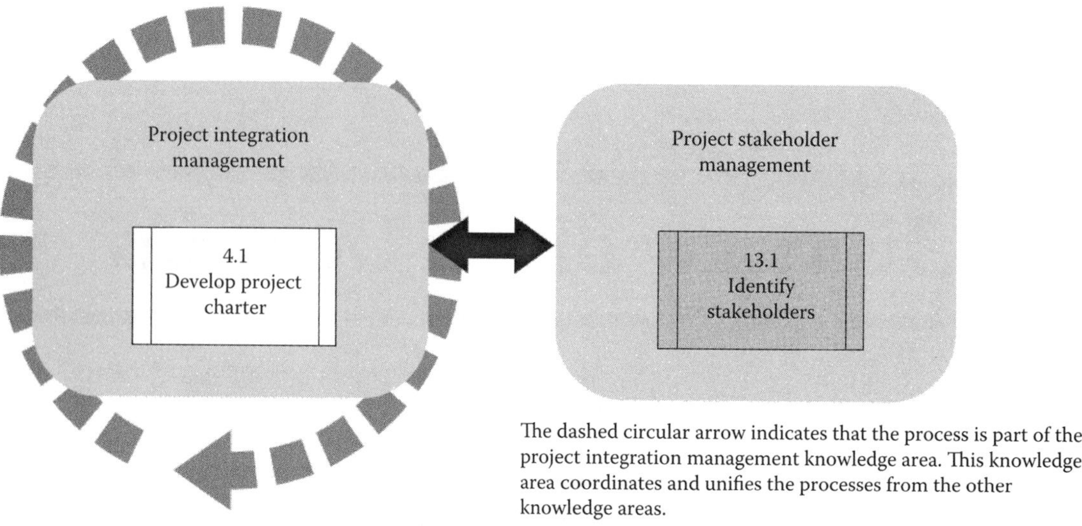

Figure 11-1 The Initiating Process Group, from Figure A1-4 in the *PMBOK® Guide, 2013*. (©PMI, 2013. Reprinted with permission.)

Study the contents of the stakeholder register and the four methods to use to classify stakeholders from the Identify Stakeholder process and refer to the list of stakeholders considered to be common to most projects in section 2.2.1 in the *PMBOK® Guide*; remember that stakeholder identification is ongoing throughout the project. Involvement of customers, sponsors, and other stakeholders from the beginning creates a shared understanding of the purpose of the project and how it fits into the organization's strategic priorities. This early involvement facilitates later deliverable acceptance and increases customer and stakeholder satisfaction.

The ECO contains eight tasks involving Initiating the Project. While many of the questions will relate to the development of the project charter, Task 5, other tasks concern stakeholders. New to this PMP® exam based on this ECO are two tasks. The first of these is:

- Task 7 – "Conduct benefit analysis with stakeholders (including sponsor, customer, subject matter expert) in order to validate project alignment with organizational strategy and expected business value" (PMI®, 2015, p. 5).

Benefit analysis is not a concept discussed in depth in the *PMBOK® Guide*; however, the benefit of each process is noted. As described in *The Standard for Program Management*, Third Edition, page 165, programs and projects deliver benefits, and a benefit is "an outcome of behaviors, products, or services that provide utility to the sponsoring organization as well as to its intended beneficiaries." Benefits are delivered by enhancing current capabilities or developing new capabilities to support the organization's strategic goals and objectives. These benefits may be quantitative, such as return on investment, or qualitative, such as enhanced employee morale. A benefits realization plan then is useful to identify and define each benefit, determine when it will be realized and how it will be sustained, and define key metrics to use to measure the benefits. Refer to section 1.6 in the *PMBOK® Guide* for a discussion on business value.

The second new task is:

- Task 8 – Inform stakeholder of the approved project charter, in order to ensure common understanding of the key deliverables, milestones, and their roles and responsibility." (PMI®, 2015, p. 5).

This task then links the two process groups and shows the need for continual stakeholder involvement and participation as appropriate in the project. See Figure 2-7 in the *PMBOK® Guide* for the relationship between the project and its stakeholders.

Planning the Project

The Planning Process Group questions cover processes to refine the project's scope and its objectives and define what must be done to attain the project's objectives. According to the ECO, they represent 24% of the questions on the PMP® exam; however, they represent 24 processes and are in all 10 of the knowledge areas in the *PMBOK® Guide*, which means there is a wealth of information to use to derive these questions. It also means you have a lot to study. Refer to Figure 11-2 in this chapter.

Although these processes culminate with the project management plan, you need to know what happens in the various knowledge areas and have a general understanding of the contents of the subsidiary plans. A number of concepts are in these various process groups, with numerous tools and techniques to create the plans and other outputs of each process. Also, the plans and other documents that are created are considered by PMI to be "living documents," meaning that as one plan is prepared, its preparation may result in the need to update other plans. The plans tend to be what is called by PMI® as progressively elaborated, meaning they are refined over time. Planning and documentation then are iterative and ongoing activities.

While it can be easy to create these plans, the importance of stakeholder involvement and buy-in cannot be overlooked, and the project manager and his or her team need to encourage stakeholder participation. To see the level of interaction of the Planning Process Group with the other process groups in a typical project, refer to Figure 3-2 in the *PMBOK® Guide*. It provides the Executing Process Group with the project management plan and other documents. It is necessary for overall project success to engage stakeholders continually to set the stage for achieving the project's objective.

Consider an area such as risks. While high-level risks are in the project charter, some risks may affect a project immediately, while in another project, time is required before risks have a significant impact. As another example, during

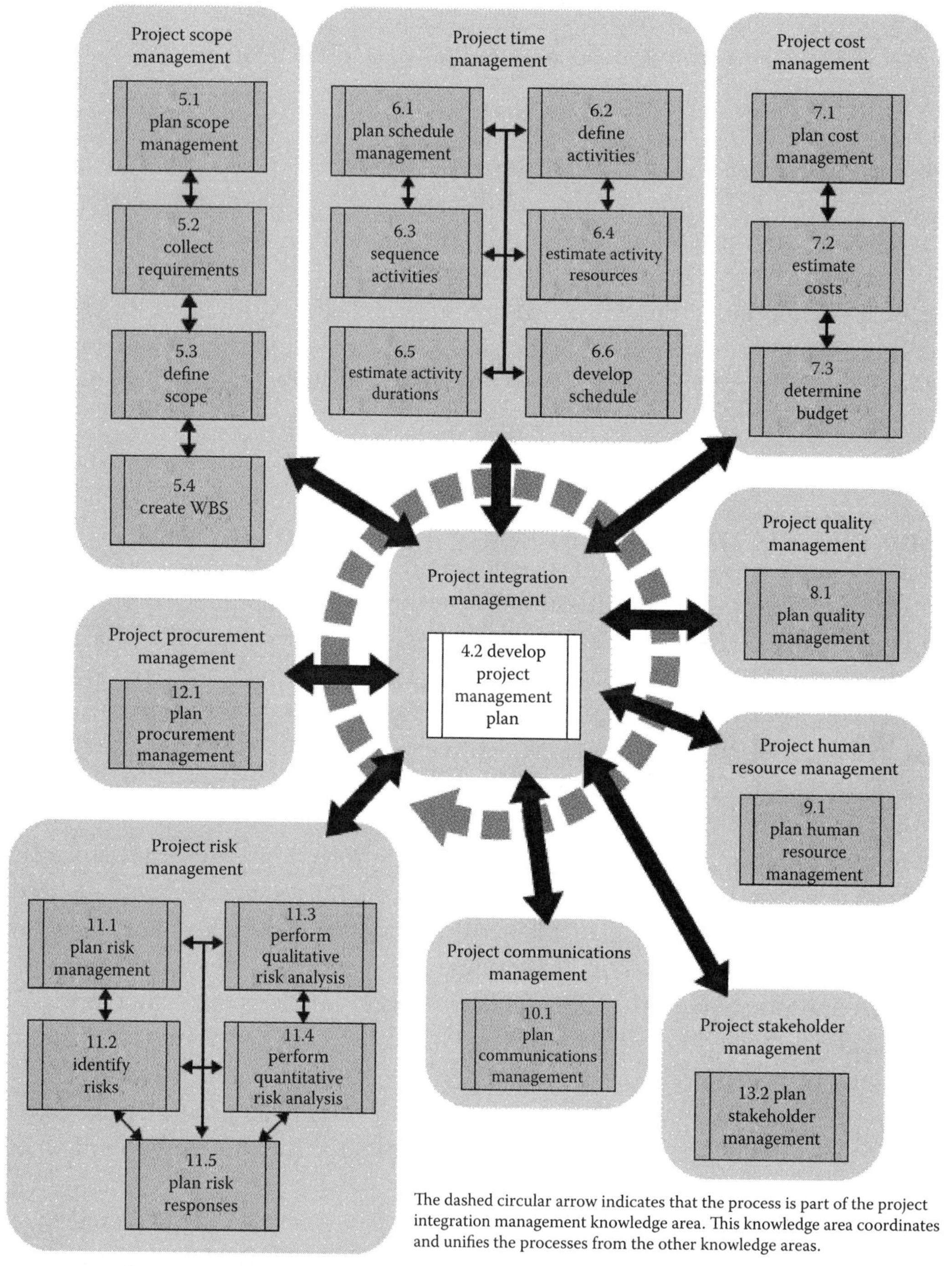

Figure 11-2 The Planning Process Group, from Figure A1-7 in the *PMBOK® Guide, 2013.* (©PMI, 2013. Reprinted with permission.)

this process, it may be noted that cost and schedule objectives may be overly aggressive. These examples show the magnitude of items to study in this process group.

The ECO has 13 tasks in this domain. Task 1 focuses on establishing detailed project deliverables. Tasks 2 through 10 and Task 13 involve the key plans that are prepared. In Task 11, the project management plan is presented to stakeholders for approval to proceed to executing. Review Section 2.2.2 in the *PMBOK® Guide* for more information on project governance.

Task 12 involves the kickoff meeting. It has been covered in previous editions of the *PMBOK® Guide* and is a best practice to follow, but it is not covered in the current edition. Therefore, take a look at our reference list. A test question may be on the kickoff meeting so you can recognize its importance regardless if your team is virtual or colocated and regardless if your team members have worked together previously.

Task 13 in this new ECO involves the stakeholder management plan, so you should study how to best prepare it and know its contents. It relates to Task 11. Remember, stakeholder involvement is essential for project success.

Executing the Project

In the Executing Process Group, the majority of the work is done on the project, and it involves 31% of the questions on the examination, as noted by the ECO. Eight of the processes and six of the knowledge areas are covered. These processes are ones required to complete the work described in the project management plan, which means it emphasizes coordinating with people and other resources, as well as managing stakeholder expectations. See Figure 11-3 in this chapter.

The key process in the figure is Direct and Manage Project Work. Typical cost and staffing levels are shown in Figure 2-8 in the *PMBOK® Guide*, and it is easy to see why this process is so important.

However, for success in all of these processes, as well as all of the other processes, interpersonal skills of the project manager cannot be underestimated. Review section 1.7 in the *PMBOK® Guide* along with Appendix X3 on interpersonal skills and expect questions on how these skills are best used in different areas. They are especially important here. Three of the Project Human Resource Management processes, one process in Project Communications Management, and one process in Project Stakeholder Management are included in this domain.

The ECO has seven tasks involving Executing the Project. Task 1 involves following the human resource management plan, which leads to the other three human resource management processes in this knowledge area that involve the project team. It also involves following the procurement management plan. The exam could include questions from the Conduct Procurement process given the work done by sellers and the increased use of outsourcing on projects. Task 2 focuses on the project management plan but also mentions leading and developing the project team

The Perform Quality Assurance process is in Executing, and it is the focus on Task 3. Recognize the importance of inspections and audits and the various tools and techniques to use. The exam could include questions from Conduct Procurements because high quality work from sellers is essential.

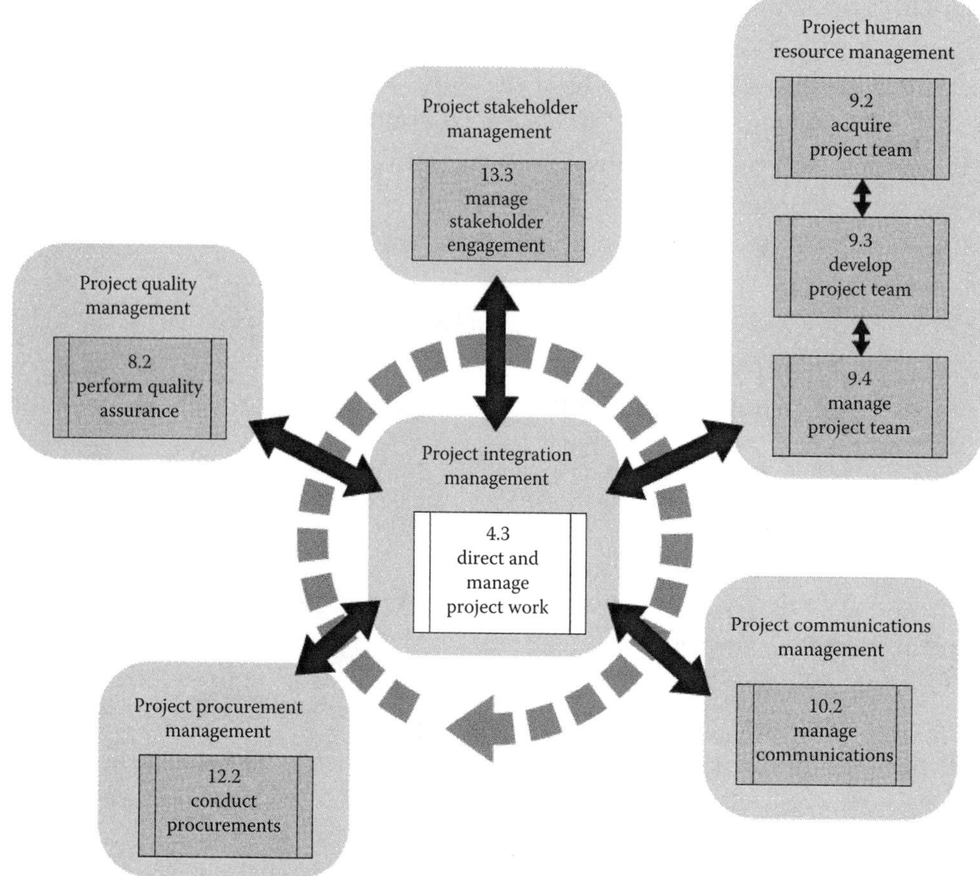

Figure 11-3 The Executing Process Group, from Figure A1-32 in the *PMBOK® Guide, 2013*. (©PMI, 2013. Reprinted with permission.)

As projects represent change, changes occur throughout the project, regardless of the life cycle used. Refer to Section 2.4 in the *PMBOK® Guide* for a discussion on the project phases and life cycles. Be aware of the need in the executing process to implement those change requests that have been approved officially, as stated in Task 4. This task supports Task 5, which notes risks may have impacts that can be negative or positive.

New are the following tasks:

- Task 6 – "Manage the flow of information by following the communications plan in order to keep stakeholders engaged an informed" (PMI®, 2015, p. 8)
- Task 7 – "Maintain stakeholder relationships by following the stakeholder management plan, in order to receive continued support and manage expectations" (PMI®, 2015, p. 8).

Both tasks build on the plans in their knowledge areas, but Task 6 focuses on the Manage Communications process. It involves more than just distributing information because it seeks to ensure stakeholders receive the information they need and understood it. Techniques to do so are on pages 300 and 301 of the *PMBOK® Guide*. Of importance is this process's benefit that enables an effective and efficient communications flow between stakeholders.

Task 7 moves to the Manage Stakeholder Engagement process and focuses on ensuring stakeholders are engaged at the right times in the project, managing expectations through negotiation and communications skills, anticipating future problems by addressing concerns that are not yet issues, and clarifying and resolving issues that have been identified. The benefit is for the project manager to increase support and minimize resistance from stakeholders, which in turn leads to a greater likelihood of project success.

Monitoring and Controlling the Project

The monitoring and controlling processes are not done in sequence after Executing, but rather they are performed from the time the project begins as shown in Figure 3-2 in the *PMBOK® Guide*. This continuous monitoring and controlling across the project's life cycle coordinates the various project phases. Corrective and preventive actions and other techniques bring the project into conformance with the project management plan. It requires tradeoffs between processes at times. Monitoring and controlling represent 25% of the questions on the PMP® exam. See Figure 11-4 in this chapter, which has 11 processes in eight knowledge areas. At the center are the Monitor and Control Project Work and Perform Integrated Change Control processes.

While monitoring and controlling implies compliance, the processes try to take a proactive approach to influencing changes that could occur and to address them as quickly as possible following agreed-upon change and configuration management procedures, which ensure only approved changes are implemented. It often leads to the need to update the various plans and other documents.

There are seven tasks in the ECO relating to monitoring and controlling. Task 1 focuses on overall monitoring of project performance by identifying variances and taking corrective actions. Therefore, this task highlights the importance of knowing the various earned value management formulas and why each one is important, as well as the purpose of other types of analysis such as variance and trend analyses.

Task 2 focuses on managing changes. It involves a change management plan, which is not described *per se* in the *PMBOK® Guide*; however, the *PMBOK® Guide* describes the change management process and the configuration processes. As you study, note how they differ and complement one another and what happens in each one.

Task 3 emphasizes quality, meaning you should study the Control Quality process and its tools and techniques. You also should note the differences and similarities between this process and the Validate Scope process.

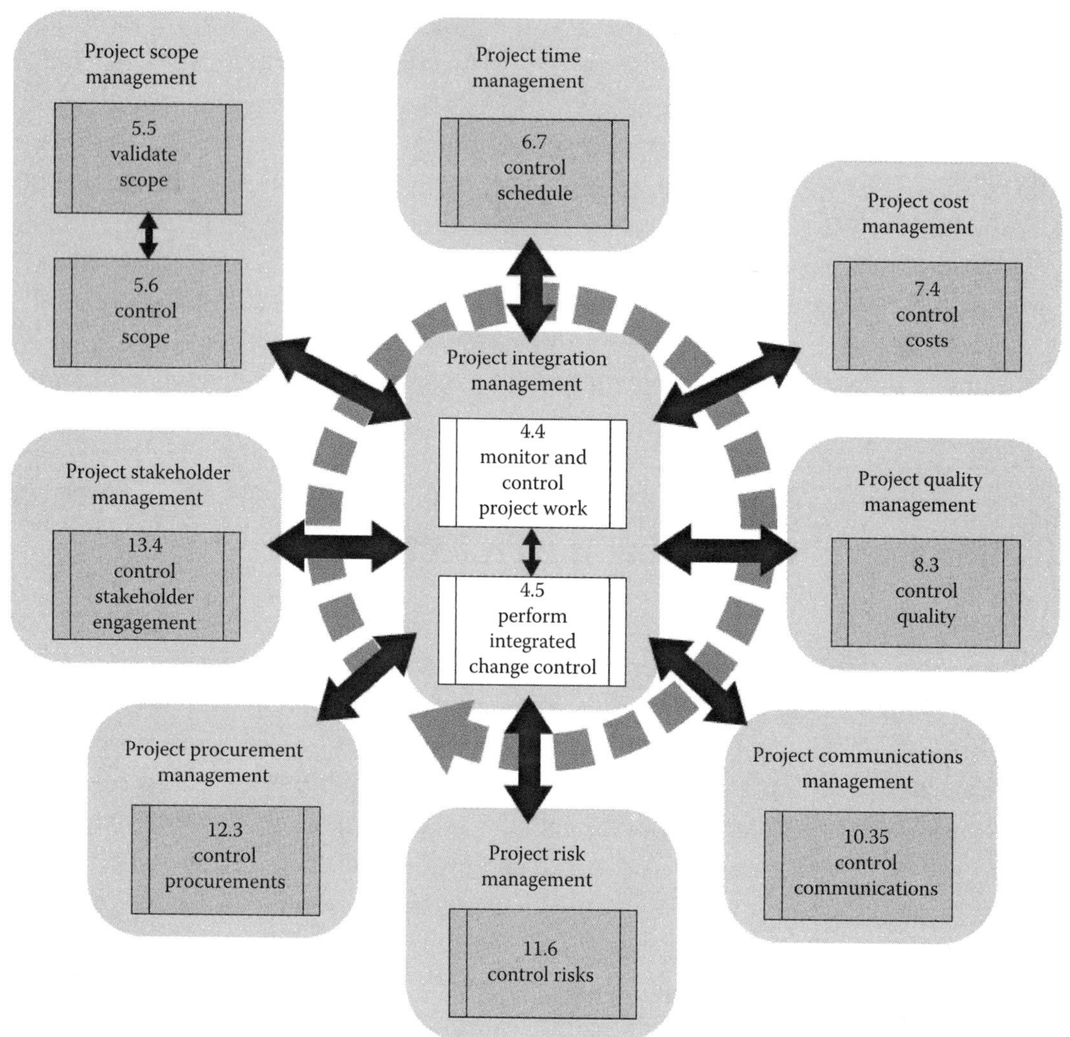

Figure 11-4 The Monitoring and Controlling Process Group, from Figure A1-41 in the *PMBOK® Guide, 2013*. (©PMI, 2013. Reprinted with permission.)

Task 4 is about monitoring and controlling risks, which means you should know when to use different tools and techniques in this process. The goal is to be proactive and not enable risks to occur without response plans in place. This leads to Task 5, reviewing the issue log. The issue log is covered briefly in the *PMBOK® Guide*. Maintaining such a log is a best practice, and risks that materialize and other concerns comprise this log. Further it is a document to update, as noted in the Control Communications process.

Tasks 6 and 7 are new and are as follows:

- Task 6 – "Capture, analyze and manage lessons learned using lessons learned management techniques in order to enable continuous improvement" (PMI®, 2015, p. 9). This task shows how monitoring and controlling processes permeate the life cycle. Before a project ends, lessons learned are collected during the project so they can be used by the project team, as well as by other projects in the organization. They are often part of organizational process assets to update in many of the process groups.

- Task 7 – "Monitor procurement activities according to the procurement plan, in order to verify compliance with project objectives" (PMI®, 2015, p. 9). This task reflects the increased outsourcing of all types of project resource and to meet the needs of projects, as well as satisfying both internal and external stakeholders, especially the sponsor and the customer. Review the Control Procurements process to familiarize yourself with its various tools and techniques. Note as well how it interacts with other processes as described on page 381 in the *PMBOK® Guide.*

Closing the Project

Although closing the project represents only seven percent of the test questions on the PMP® exam, and has only two processes from two knowledge areas, its importance cannot be overlooked. When the process is complete, the project or a phase is closed, and it indicates that all the 47 defined processes in the *PMBOK® Guide* are complete. It also addresses any premature closure and why it occurred.

In the ECO, it has seven tasks, and there are no new tasks in this June 2015 update.

Task 1 involves obtaining final acceptance from stakeholders that the scope and the deliverables of the project were achieved. This task notes the importance of early and ongoing engagement of stakeholders throughout the project.

Task 2 emphasizes transferring ownership of the deliverables to stakeholders (e.g., an operations group, a functional group, a customer, and end users). An orderly transition and stakeholder involvement is needed. These stakeholders must have early involvement, recognize their roles and responsibilities, and the need for them to be prepared to receive the product, service, or result and sustain its benefits.

Task 3 involves financial, legal, and administrative closure. Review the steps involved in administrative closure from the Close Project or Phase process. It also emphasizes communications to stakeholders about formal project closure.

Task 4 involves preparing and sharing the final project report. This report should be noted in the communications management plan and describes project successes and areas for improvement. Task 5 continues emphasizing collecting the project's lessons learned and conducting a comprehensive project review. This is the time when the project's lessons learned are reviewed to see if all should be retained or if some are no longer valuable. The final review includes the key project stakeholders to further discuss the project and add to the lessons learned. This leads to Task 6, which involves archiving the project's documents and materials. This task is significant for future projects, which should have easy access to the lessons learned and be able to tailor them.

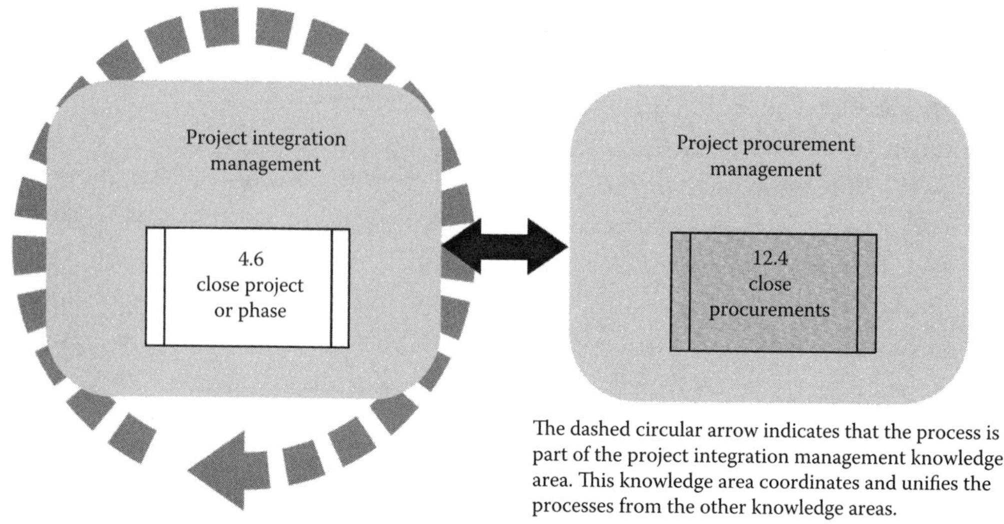

Figure 11-5 The Closing Process Group, from Figure A1-53 in the *PMBOK® Guide, 2013*. (©PMI, 2013. Reprinted with permission.)

Task 8 involves feedback from stakeholders. This task may be performed by someone external to the project and may involve interviews or surveys. The goal is to show stakeholders the importance of their involvement on the project and to actively listen to learn what could be improved in the future. It follows information from the stakeholder management plan. Refer to Figure 11-5.

Practice Test

This practice test is designed to simulate PMI®'s 200-question PMP® certification exam. It is our recommendation that you first try this test online to see your areas of strength and areas for improvement. Then, take the test which follows so you can see the rationale for the right answer. You can take both versions of this test as often as you wish to do so.

INSTRUCTIONS: Note the most suitable answer for each multiple-choice question in the appropriate space on the answer sheet.

1. You recently took over a relatively new project expected to last another seven years. The previous project manager completed some of the WBS. When you begin to define the project activities, you realize that the WBS work packages expected to occur in the next year are planned in detail, but the work packages for later in the future (three years or more) are not planned with much detail, if any detail at all. You determine—

   a. It is a major problem. The WBS is incomplete and you need to redefine the project scope to complete the project schedule.
   b. It is a problem that must be resolved quickly. The previous project manager was not done with the WBS, and you must stop the project to complete the WBS in sufficient detail.
   c. It is not a problem at this time. The previous project manager was using the rolling wave planning technique, so you are able to continue defining the activities.
   d. It is not a problem at this time. You can only plan what you know. You plan to communicate to the project sponsor that the WBS is not sufficient to plan the whole project and that the sponsor can worry about the details.

2. You are managing a project to transform your organization into one that is more customer centric as your executives realize they no longer can assume existing customers will stay with your firm given offerings by the competition. You and your team identified over 300 stakeholders and classified them. You realize 45 are influential but negative to the project. You then prepared your stakeholder management plan. Now, you find you are addressing concerns from these stakeholders plus others, and it consumes most of your time. However, it is important because—

   a. They have concerns over priorities and procedures
   b. These concerns may become issues
   c. You need committed stakeholders to ensure your project continues
   d. You realize your project has the greatest chance for success if it operates in a projectized environment, and stakeholder support is paramount

3. On your project to construct a new runway for your City's airport, you are in the process of selecting vendors for various parts of this project. You have conducted your make-or-buy analysis and have issued Requests for Proposals. You believe it is important to examine the readiness of the vendor to provide the desired end state. This means you are using—

   a. Proposal evaluation techniques
   b. Multi-disciplinary review teams
   c. Analytical techniques
   d. Independent estimates

4. Requirements typically are classified into product requirements and project requirements. Capturing and managing both types of requirements is important for project success, so you and your team decided to follow this classification system on your project to modernize all the telecommunications equipment in your company. During such an approach, all the following are examples of solution requirements EXCEPT—
   a. Action requirements
   b. Level of service requirements
   c. Security requirements
   d. Performance requirements

5. Change control procedures, configuration management knowledge base, versions, and baselines in the Develop Project Management Plan process are:
   a. Enterprise environmental factors
   b. Organizational process assets
   c. Part of the project's configuration management plan, which as a subsidiary plan will be part of the project management plan
   d. Part of the organization's management practices

6. You are managing a project that has five subcontractors. You must monitor contract performance, make payments, and manage provider interfaces. One subcontractor submitted a change request to expand the scope of its work. You decided to award a contract modification based on a review of this request. All these activities are part of—
   a. Control Procurements
   b. Conduct Procurements
   c. Form Contract
   d. Configuration Management

7. Your project management plan has been approved, and since your company follows a stage-gate approach, you are now in the executing phase. You have collected a lot of data, and these data are viewed as—
   a. Ones set forth in the PMIS
   b. The lowest level of detail to derive information
   c. Ones recommended through a survey of your stakeholders to assess their key communications requirements
   d. A major part of the project management plan

8. The performance measurement baseline consists of all the following EXCEPT—
   a. Scope baseline
   b. Requirements baseline
   c. Schedule baseline
   d. Cost baseline

9. While working as the project manager on a new project to improve overall ease of use in the development of a railroad switching station, you have decided to add a subject matter expert who specializes in ergonomics to your team. She has decided to observe the existing approach as you and your team work to define requirements for the new system. This method is also called—

   a. Mentoring
   b. Coaching
   c. Job shadowing
   d. User experimentation

10. In addition to providing support to the project, quality assurance also provides an umbrella for—

    a. Plan-do-check-act
    b. Improvement of quality processes
    c. Project management maturity
    d. Work performance information

11. As you manage the railroad switching station project, you are concerned that the business analyst who was responsible for preparing the WBS may have overlooked some parts of the project. In order to see if the WBS requires enhancements you decide to—

    a. Perform a cause-and-effect diagram
    b. Meet with your sponsor
    c. Use an affinity diagram
    d. Review the accompanying WBS Dictionary with a member of the PMO

12. Assume that you are managing a project in which 80% has been outsourced to various sellers. In terms of contract administration, a key aspect is

    a. Having a contract administrator assigned to your core team and reports to you
    b. Ensuring there are few, if any, claims
    c. Considering the sellers as members of the project team
    d. Managing the interfaces among the sellers

13. If you apply the configuration management system along with change control processes project wide, you will achieve all but one of the which following objectives?

    a. Provide the basis for which the product configuration is defining
    b. List the approved configuration identification
    c. Document the specific responsibilities of each stakeholder in the Perform Integrated Change Control process
    d. Ensure the composition of a project's configuration items is correct

14. You need to outsource the testing function of your project. Your subcontracts department informed you that the following document must be prepared before conducting the procurement:

    a. Make-or-buy analysis
    b. Procurement management plan
    c. Evaluation methodology
    d. Contract terms and conditions

15. Constraints common to projects include—

    a. Scope, quality, schedule, budget, and risk
    b. Scope, teaming, planning, and resources
    c. Scope
    d. Resources and communication

16. You are developing a project charter, and it requires first that a business case be prepared to make a decision to go ahead with a project Your project is to lessen the environmental impact of a landfill so its business case is created primarily as a result of a(n)—

    a. Sustainability impact
    b. Customer request
    c. Social need
    d. Ecological impact

17. To identify inefficient and ineffective policies, processes, and procedures in use on a project, you should conduct—

    a. An inspection
    b. A process analysis
    c. Benchmarking
    d. A quality audit

18. Your project management office implemented a project management methodology that emphasizes the importance of integrated change control. It states that change requests may include—

    a. Indirect changes
    b. Defect repairs
    c. Informal requests from executives
    d. Ones that do not impact baselines

19. Since projects involve change, change requests then will be needed. Every documented change request must be approved by a responsible person, who is—

    a. The project manager or the sponsor
    b. A member of the CCB
    c. Identified in the change management plan
    d. Identified in the project management plan

20. A number of tools and techniques are helpful in the Perform Integrated Change Control process. If you want to implement an integrated change control process, you should use—

    a. Configuration management software
    b. A project management information system
    c. Project status review meetings
    d. Change control meetings

21. Having worked previously as a software project manager, you were pleased to be appointed as the project manager for a new systems integration project designed to replace the existing air traffic control system in your country. You found a requirements traceability matrix to be helpful on software projects, so you decided to use it on this systems integration project. Using such a matrix helps to ensure that each requirement—

    a. Adds quality and supports the organization's quality policy
    b. Adds business value as it links to business and project objectives
    c. Sets forth the level of service, performance, safety, security, and compliance
    d. Shows the impact to other organizational areas and to entities outside of the performing organization

22. You are working on a complex project in the medical device field. You have about 55 internal stakeholders and a very large number of external stakeholders including regulatory and consumer interest groups along with the medical profession. You need to be aware of their risk tolerances as they are examples in the Direct and Manage Project Work process of—

    a. The need to review the stakeholder management plan early in the process
    b. Enterprise environmental factors
    c. Organizational process assets
    d. The importance of classifying stakeholders as to interest and influence

23. Which of the following is focused on the specification of both the deliverables and the processes as part of the Perform Integrated Change Control process?

    a. Scope change control system
    b. Configuration control
    c. Change control board
    d. Configuration status audits

24. Which of the following tools is used in process analysis to determine the underlying causes of defects?

    a. Root cause analysis
    b. Assumptions analysis
    c. Cost-benefit analysis
    d. Quality metrics

25. All of the following statements concerning Validate Scope and Control Quality are true EXCEPT—

    a. The processes can be performed in parallel
    b. Both processes use inspection as a tool and technique
    c. Validate Scope is concerned with the acceptance of deliverables, and Control Quality is concerned with meeting quality requirements for the deliverables
    d. Validate Scope typically precedes Control Quality

26. There are a number of activities that must be done to close a project or a phase. They are step-by-step methodologies. The first one addresses—

    a. Transitioning the product or service to end users, the customer, or an operations group
    b. Measuring the completeness of the project's deliverables against the requirements
    c. Making sure all the activities necessary to satisfy exit criteria for a phase or the entire project are followed
    d. Holding a lessons learned session with interested stakeholders

27. You are in the process of performing quality assurance on your product and find that some requirements are not as complete as they should be, which causes rework and adds costs to your overall project. One approach to determine your next steps is—

    a. Determining the cost of quality
    b. Using a checklist
    c. Identifying the key issues and any suitable alternatives
    d. Comparing the cost of rework to the life-cycle costs of the project

28. The project scope baseline should be used in the Identify Risks process because it—

    a. Identifies project assumptions
    b. Identifies all work that must be done; therefore, it includes all risks on the project
    c. Helps organize all work that must be done on the project
    d. Contains information on risks from prior projects

29. Although there are various tools and techniques to consider as you collect requirements on your project, one approach that supports the concept of progressive elaboration is—

    a. Idea/mind mapping
    b. Affinity diagrams
    c. Prototypes
    d. Joint Application Design® sessions

30. Data from which of the following are used to Perform Quality Assurance?

    a. Control Quality and Plan Quality Management
    b. Performance Reporting
    c. Variance analysis
    d. Tools from Direct and Manage Project execution

31. An approach to provide insight into the health of the project and to identify any areas that require special attention is to—

    a. Conduct periodic status reviews
    b. Prepare regular status and progress reports
    c. Prepare forecasts of the project's future
    d. Continuously monitor the project

32. Although your company's project life cycle does not mandate when a project review should be conducted, you believe it is important to review performance at the conclusion of each phase. The objective of such a review is to—

    a. Determine how many resources are required to complete the project according to the project baseline
    b. Adjust the schedule and cost baselines based on past performance
    c. Obtain customer acceptance of project deliverables
    d. Determine whether the project should continue to the next phase

33. The key management skills required during the adjourning stage of team development include all but which one of the following?

    a. Evaluating
    b. Reviewing
    c. Celebrating
    d. Improving

34. Assume that your actual costs are $1,000; your planned value is $1,200; and your earned value is $1,500. Based on these data, what can be determined regarding your schedule variance?

    a. At –$300, the physical progress is being accomplished at a slower rate than is planned, indicating an unfavorable situation.
    b. At +$300, the situation is favorable, as physical progress is being accomplished ahead of your plan.
    c. At +$500, the situation is favorable, as physical progress is being accomplished at a lower cost than was forecasted.
    d. At –$300, you have a behind-schedule condition, and your critical path has slipped.

35. The key to effective cost control is—

    a. Using earned value to forecast project status
    b. Focusing on projected expenditures and actively networking with key stakeholders to ensure funds will be available as requested
    c. Informing stakeholders of the project's cost status
    d. Managing the approved cost baseline and any changes to it

36. The CPI on your project is 0.44, which means that you should—

    a. Place emphasis on improving the timeliness of the physical progress
    b. Reassess the life-cycle costs of your product, including the length of the life-cycle phase
    c. Place emphasis on improving the productivity by which work was being performed
    d. Recognize that your original estimates were fundamentally flawed, and your project is in an atypical situation

37. Project deliverables are the outputs that include the product, service, or result of the project as well as ancillary results. These ancillary results should be in the—

    a. Requirements management plan
    b. Scope management plan
    c. Project scope statement
    d. Project acceptance criteria

38. Which of the following tools and techniques is used in the Close Project or Phase process?

    a. Project management methodology
    b. Work performance information
    c. Expert judgment
    d. Project management information system

39. After the project scope statement is complete, it may be necessary to update other project documents. All the following are examples of a document that may require updates EXCEPT—

    a. Project charter
    b. Stakeholder register
    c. Requirements documentation
    d. Requirements traceability matrix

40. Your company, noted for its use of innovative technology in its work, is also noted for exceeding its budget. On your project, you have been asked by the executive team to keep costs under control. You are focusing your attention on—

    a. Informing your stakeholders of approved change requests and their costs
    b. Using earned value and following the 50/50 rule
    c. Focusing on work performance information as you begin your work to control your project costs
    d. Involving stakeholders to ensure change requests are decided as quickly as possible

41. While managing a large project in your organization, you realize that your project team requires training in contract administration because you will be awarding several major subcontracts. After you analyze your project requirements and assess the expertise of your team members, you decide that your team will need a one-week class in contract administration. This training should—

    a. Commence as scheduled and stated in the staffing management plan
    b. Commence as scheduled and stated as part of the procurement management plan
    c. Be scheduled if necessary after performance assessments are prepared and after each team member has had an opportunity to serve in the contract administrator role
    d. Commence as scheduled and stated in the team development plan

42. Assume that on your project, you are using earned value management. You also are conducting performance reviews. You recognize, though, that an important aspect of cost control is determining the cause and variance compared to the cost baseline. You have found that—

    a. The most frequently analyzed measurements are CPI and SPI
    b. The milestone method is the most accurate way to assess performance using earned value
    c. Graphical techniques will enhance how the information is received
    d. The range of acceptable variances will tend to decrease over time

43. Your project sponsor has asked you, "What do we now expect the total job to cost?" Given that you are using earned value, you should calculate the—

    a. To-complete performance index
    b. Estimate to complete
    c. Estimate at completion
    d. Budget at completion

44. One key reason that the Develop Project Charter process is so important is that it—

    a. Documents the boundaries of the project
    b. States the methods for acceptance of the project's deliverables
    c. Describes the project's characteristics
    d. Links the project to the ongoing work of the organization

45. Your company has been awarded a contract for project management consulting services for a major government agency. You were a member of the proposal writing team, are PMP® certified, and you are the project manager. You are now working to prepare your project management plan, which is to be submitted in one week. You decided to use some facilitation techniques to help develop your plan. While a number are possible, you selected—

    a. Meeting management
    b. Checklist analysis
    c. SWOT analysis
    d. Assumptions analysis

46. Assume you had a phase gate meeting with your Governance Board for your project to develop the next generation radar system as part of the nation's airspace modernization program. At this meeting, the Board approved your project management plan. However, as you begin to execute your plan, an organizational process asset to consider is—

    a. Stakeholder risk tolerances
    b. The organization's culture
    c. Hiring and firing guidelines
    d. Process measurement data base

47. Consider the data in the table below. Assume that your project consists only of these three activities. Your estimate at completion is $4,400.00. This means you are calculating your EAC by using which of the following formulas?

Activity	*% Complete*	*PV*	*EV*	*AC*
A	100	2,000	2,000	2,200
B	50	1,000	500	700
C	0	1,000	0	0

    a. EAC = AC/EV × BAC
    b. EAC = AC/EV × [work completed and in progress] + [actual or revised cost of work packages that have not started]
    c. EAC = AC + Bottom-up ETC
    d. EAC = % complete × BAC

48. Rolling wave planning in the Create WBS process refers to situations in which—

    a. Certain deliverables or subprojects will be accomplished far into the future
    b. Additional work is added to the project after the scope baseline has been established; therefore, additional decomposition is required
    c. Identification codes for the WBS elements cannot be determined until the schedule activity list is complete in case revisions are required
    d. Subprojects are developed by external organizations and then become part of the WBS for the entire project

49. The lessons learned documentation is an output from the—

    a. Identify Stakeholders process
    b. Develop Project Management Plan process
    c. Manage Communications process
    d. Plan Communications Management process

50. Your experience has taught you that inappropriate responses to cost variances can produce quality or schedule problems or unacceptable project risk. When leading a team meeting to discuss the importance of cost control, you note that cost control is concerned with—

    a. Influencing the factors that create change to the authorized cost baseline
    b. Developing an approximation of the costs of the resources needed to complete the project
    c. Allocating the overall cost estimate to individual work items
    d. Establishing a cost performance baseline

51. You are pleased to be the project manager for a new video conferencing system for your global organization. You want it to be one that is easy to use and is state of the art. As the project manager, you also are the project leader. You realize leadership is critical throughout the phases of the project and its key elements are—

    a. Respect and trust
    b. Political and cultural awareness
    c. Negotiation and influencing
    d. Decision making and conflict management

52. The WBS represents all product and project work, including project management. It is sometimes called the—

    a. Control account level
    b. 100% rule
    c. Integration of scope, cost, and schedule for comparison to the earned value
    d. The code of accounts

53. Your company is in the project management training business. In addition, the company publishes several exam study aids for the PMP® and CAPM® exam. You have your PMP®, and you have been appointed as the project manager to make sure your company's training materials are updated to be aligned with the new *PMBOK® Guide*. You must complete your project in six months. You are now in month four. Many of your team members have been working on other projects as the company uses matrix management. In a performance review meeting today, you informed your Governance Board that you did not think you could complete this project in the remaining two months. You were informed that additional resources were not available, but you had to complete your project on time. Your best course of action is to—

    a. Revise your schedule baseline
    b. Use fast tracking
    c. Adjust leads and lags now in your schedule
    d. Use modeling techniques

54. You are trying to determine whether or not to conduct 100% final system tests of 500 ground-based radar units at the factory. The historical radar field failure rate is 4%; the cost to test each unit in the factory is $10,000; the cost to reassemble each passed unit after the factory test is $2,000; the cost to repair and reassemble each failed unit after factory test is $23,000; and the cost to repair and reinstall each failed unit in the field is $350,000. Using decision tree analysis, what is the expected value if you decide to conduct these tests?

    a. $5.5 million
    b. $5.96 million
    c. $6.42 million
    d. $7 million

55. Motivation is dynamic and complex. The overall success of the project depends on the team's commitment to it, which is directly related to motivation. On your project, you want to create an environment to meet project objectives, but you want to motivate your team members by providing them with what they value the most. An example of a value is—

    a. Clarifying why you made a certain decision
    b. Using active and passive listening techniques
    c. Providing a high quality of information exchange
    d. Promoting job satisfaction

56. Each time you meet with your project sponsor, she emphasizes the need for cost control. To address her concerns, you should provide—

    a. Work performance information
    b. Cost baseline updates
    c. Resource productivity analyses
    d. Trend analysis statistics

57. One output of the Control Costs process is cost forecasts, which is when—

    a. Modifications are made to the cost information used to manage the project and are communicated to stakeholders
    b. Trend analyses are performed and communicated to stakeholders
    c. A budget update is required and communicated to all stakeholders
    d. A calculated EAC value or a bottom-up EAC value is documented and communicated to stakeholders

58. You work for an electrical utility company and will be managing a project to build a new substation that will serve a new industrial park. This project was authorized because of a—

    a. Business need
    b. Market demand
    c. Technological advance
    d. Customer request

59. You are about to close your project. In doing so you are using trend analysis, which is—

    a. An analytical technique
    b. A way to show corrective actions taken on the project
    c. A method of assessing the usefulness of causal analysis
    d. An approach that also involves expert judgment

60. At the time the risk register is first prepared, it should contain—

    a. Root causes of risk
    b. Risk owner
    c. List of risks requiring near-term responses
    d. List of potential responses

61. You are following a collaborative approach as you lead you team and are conducting a team performance assessment in addition to individual assessments. The performance of a successful team is measured in terms of—

    a. Project commitment
    b. Technical success
    c. Enhanced ability to work as a team
    d. Effective decision making

62. Your project is considered very risky. You plan to perform numerous what-if scenarios on your schedule using simulation software that will define each schedule activity and calculate a range of possible durations for each activity. The simulation then will use the collected data from each activity to calculate a distribution curve (or range) for the possible outcomes of the total project. Your planned approach is an example of which following technique?

    a. PERT
    b. Monte Carlo analysis
    c. Linear programming
    d. Concurrent engineering

63. Project execution must be compared, and deviations must be measured for management control according to the—

    a. Scope baseline
    b. Performance measurement baseline
    c. Schedule baseline
    d. Control system

64. A number of items may be part of the schedule data for the project. The amount of additional detail will vary, but the data should include all the following items EXCEPT—

    a. Schedule activities
    b. Activity attributes
    c. Identified assumptions
    d. Resource breakdown structure

65. Lessons learned are important throughout project management. Assume you have documented them as you updated the organizational process assets in the Manage Stakeholder Engagement process. An example of one to include in this process is—

    a. Root causes of issues
    b. Effective conflict management approaches
    c. Meeting minutes
    d. Risk log

66. Which tool or technique is NOT used for Control Schedule control?

    a. Performance reviews
    b. Project management software
    c. Work performance information
    d. Leads and lags

67. As stakeholders engage with the project, the overall effectiveness of the stakeholder management strategy can be evaluated. As the project manager, this means you should—

    a. Revise the stakeholder matrix
    b. Review the roles of the stakeholders
    c. Update the project management plan
    d. Update the stakeholder engagement plan

68. Recording and reporting information regarding when appropriate configuration information should be provided and regarding the status of proposed and approved changes effectively is done through—

    a. Configuration status accounting
    b. Configuration verification and audit
    c. Project management methodology
    d. A project management information system (PMIS)

69. Decomposition is a technique used to break larger, complex items into smaller and more manageable items. Which following statement best describes the role decomposition plays in creating the WBS?

    a. Final output of creating the WBS is described in terms of phases of a project life cycle.
    b. Final output of creating the WBS is described in terms of schedule activities.
    c. Final output of creating the WBS is described in terms of verifiable products, services, or results.
    d. Final output of creating the WBS is described in terms of the scope of the project.

70. The schedule can be presented in a number of formats. Assume your stakeholders want to see it as a time-scaled logic diagram. This—

    a. Is optimized to show relationships between activities
    b. Is a pure logic diagram
    c. Is a logic bar chart
    d. Is one in which each work package is planned as a set of activities

71. Activity attributes are used to extend the description of the activity and to identify its multiple components. In the early stages of the project, an example of an activity attribute is—

    a. Activity codes
    b. Activity description
    c. Predecessor and successor activities
    d. Activity name

72. You are working on a new project in your city to construct an environmentally friendly landfill. The existing site is so undesirable that many residents have moved to other neighboring cities because of their proximity to it. However, even though the project has the support of the public, you need to have a number of hearings with the city's government before you are authorized to begin work. As you are in the planning phase of the project, you are waiting for these hearings to be scheduled and held before you can begin site preparation. These hearings are an example of—

    a. A milestone
    b. An external dependency
    c. An item to be scheduled as a fragnet
    d. A mandatory dependency

73. You are working on a project and want to know how many activities in the previous month were completed with significant variances. You should use a(n)—

    a. Control chart
    b. Inspection
    c. Scatter diagram
    d. Trend analysis

74. Your project has a budget of $1.5 million for the first year, $3 million for the second year, $2.2 million for the third year, and $800,000 for the fourth year. Most of the project budget will be spent during—

    a. Starting the project
    b. Organizing and preparing
    c. Carrying out the work
    d. Closing the project

75. If you decide to follow an open subordination approach to resolving conflict, you are using which style of conflict resolution?

    a. Avoiding
    b. Accommodating
    c. Compromising
    d. Collaborating

76. Typically, the seller receives formal written notice that the contract has been completed by the—

    a. Project manager
    b. Authorized procurement administrator
    c. Member of the project management team responsible for daily contract administration
    d. Purchasing department head

77. You are working in the Control Procurements process and your project is using five contractors. You need to update a number of organizational process assets. An example is—

    a. Procurement management plan
    b. Procurement documentation
    c. Correspondence
    d. Warranties

78. You are beginning a new project staffed with a virtual team located across five different countries. To help limit conflict and misunderstandings concerning the justification, objectives, and high-level requirements of the project among your team members and their functional managers, you ask the project sponsor to prepare a—

    a. Memo to team members informing them that they work for you now
    b. Project charter
    c. Memo to functional managers informing them that you have authority to direct their employees
    d. Human resource management plan

79. To anticipate and help develop approaches to deal with potential quality problems on your project, you want to use a variety of root-cause analysis techniques including all the following approaches EXCEPT—

    a. Fishbone diagrams
    b. Ishikawa diagrams
    c. System or process flowcharts
    d. Checklists

80. All of the following are examples of ways to generate options for mutual gain during negotiations EXCEPT—

    a. Separating inventing from deciding
    b. Options broadening
    c. Zero-sum game analysis
    d. Multiplying options by shuttling between the specific and the general

81. Recently, your company introduced a new processing system for its products. You were the project manager for this system and now have been asked to lead a team to implement needed changes to increase efficiency and productivity. To help you analyze the process outputs, you and your team have decided to use which following technique?

    a. System flowcharts
    b. Design of experiments
    c. Pareto analysis
    d. Control charts

82. As the project manager, you are negotiating with functional managers and other project managers to staff your project with the required levels of expertise. To determine the most appropriate criteria to select your team, you decided to use—

    a. Expert judgment
    b. A virtual team
    c. Set clear expectations
    d. Multi-criteria decision analysis

83. Based on quality control measurements on your manufacturing project, management realizes that immediate corrective action is required to the material requirements planning (MRP) system to minimize rework. To implement the necessary changes you should follow—

    a. The organization's quality policy
    b. The quality management plan
    c. Established operational definitions and procedures
    d. A defined integrated change control process

84. You are the project manager on a project to improve traffic flow in the company's parking garage. You decide to use flowcharting to—

    a. Help anticipate how problems occur
    b. Show dependencies between tasks
    c. Show the results of a process
    d. Forecast future outcomes

85. Recognizing the importance of people on projects, you are preparing your human resource management plan. You decide to take a look at the organizational process assets so you review—

    a. Escalation procedures for handling issues
    b. The organizational culture
    c. Personnel administration policies
    d. Existing human resources

86. Schedule control is one important way to avoid delays. While planning and executing schedule recovery, one tool available to you to control the schedule is—

    a. Changing the schedule management plan
    b. Immediately re-baselining
    c. Adjusting leads and lags
    d. Changing all project and resource calendars

87. You have been the project manager for your nuclear submarine project for four years. While you did not assume this position until the project management plan had been prepared and approved, you find you spend a significant amount of time collecting data and communicating. You also spend time reviewing the impact of project changes and implementing ones that have been approved. Often you have had to modify a non-conforming product, which means you are spending time on—

    a. Corrective actions
    b. Updating the project's requirements
    c. Updating the traceability matrix
    d. Defect repair

88. You were assigned recently as the project manager of a program management office project to implement a new enterprise-wide scheduling system for use throughout your company. You identify the need for a project charter to provide you with appropriate authority for applying resources, completing the project work, and formally initiating the project. Who should issue the project charter?

    a. The project manager—you
    b. The customer
    c. The person who formally authorizes the project
    d. A member of the training and development department as they will own the training on the new system

89. In which of the following methods of resolving conflict will the conflict typically reappear again in another form?

    a. Smoothing
    b. Compromising
    c. Collaborating
    d. Confronting

90. Statistical sampling is a method in Control Quality to determine the conformance to requirements for some component or product of a project. Its greatest advantage is that it—

    a. Does not require a large expenditure of resources
    b. Is accurate enough with a sampling of less than 1%
    c. Does not require 100% inspection of the components to achieve a satisfactory inference of the population
    d. Needs to be conducted only when a problem is discovered with the end product or when the customer has some rejects

91. Your project sponsor wants to know whether process variables are within acceptable limits. To answer this question, you should—

    a. Conduct a process analysis
    b. Conduct a root cause analysis
    c. Use a control chart
    d. Use a run chart

92. You are preparing your cost estimate for your project in robotics. Already, some competitors have learned about this project, and they want to be possible purchasers of the project when it is complete. You consider this to be an opportunity, which means—

    a. Your cost estimate's accuracy rate should be from –5% to +10%
    b. You need to reduce activity costs as much as possible
    c. Your estimate should include the cost of financing
    d. You should include indirect costs in your estimate

93. You are planning a project and want to account for how the project will be managed in the future. While building your cost performance data, you want to provide guidance for when the project is later executed, because you know that different responses are required depending upon the degree of variance from the baseline. For example, a variance of 10 percent might not require immediate action, whereas a variance of 20 percent will require more immediate action and investigation. You decide to include the details of how to manage the cost variances as part of which following plan?

    a. Cost management plan
    b. Change management plan
    c. Performance measurement plan
    d. Variance management plan

94. Assume that you are managing a project team. Your team is one in which its members confront issues rather than people, establish procedures collectively, and is team oriented. As the project manager, which of the following represents your team's stage of development and the approach you should use during this time?

    a. Storming; high directive and supportive approach
    b. Norming; high directive and low supportive approach
    c. Norming; high supportive and low directive approach
    d. Performing; low directive and supportive approach

95. You are finalizing all the contracts and ensuring that they are closed. The Close Procurements process involves all the following administrative actions EXCEPT—

    a. The procurement administrator is reassigned
    b. Finalizing open claims
    c. Updating the project records to show the final contract results
    d. Archiving the contracts and contract records for future use

96. You are working on a project and want to identify the cause of problems in a process by the shape and width of the distribution of the process variables. You should use a—

    a. Histogram
    b. Pareto chart
    c. Scatter diagram
    d. Trend analysis

97. You are working on a construction project in a city different from your headquarters' location. You and your team have not worked in this city, City B, previously, and you lack knowledge of the local building codes. You had a team member review the codes, and he said they were in far greater detail than those in your city, City A. When you asked him how much time he would need to spend to gain a complete understanding of these codes, he estimated that at least five weeks would be needed. You then decided it would be more cost effective to hire a local person from City B who specializes in this area. As a result, as you prepare your schedule and estimate your resource requirements for this project, you should coordinate this work closely with which of the following processes:

    a. Estimate Costs
    b. Define Activities
    c. Determine Budgets
    d. Develop Schedule

98. Assume that you are managing a project that once completed will take you company into new markets. Since it is so significant it has the interest of executive managers and other key internal stakeholders. You know for success, the marketing department will play a key role. You and your team identified them as key stakeholders. You met with the Chief Marketing Officer, and she indicated she would support the project. However, the Chef Marketing Officer has attended only the first status review meeting but none of the subsequent meetings. She also has not sent someone from her staff. This situation shows the importance of—

    a. Executive support
    b. The need to escalate this issue to your sponsor
    c. Maintaining the stakeholder register
    d. Engaging stakeholders at certain stages

99. The nature of project work is such that it inevitably changes. You know this is the case on your software project as you were about 50% done when the company announced all software work was to be done using agile, even work in progress, and your project was using waterfall. You now believe it is time to ensure your stakeholders have not changed, and you want to reassess them to make sure you do not have new stakeholders or have ones whose positions have changed. You decide to—

    a. Personally review the stakeholder register that was prepared when the project began
    b. Update the stakeholder engagement plan
    c. Use subject matter experts
    d. Reevaluate your power/interest grid

100. Assume you are working as the project manager on the first project in your company to use the critical chain approach to scheduling. You are a PMP® and also are certified in critical chain. You are getting ready for a performance review with your Governance Board, and you expect they will ask questions about—

    a. The magnitude of variance against the schedule baseline
    b. Schedule risk
    c. Performance to date since the past review meeting
    d. The buffer needed and buffer remaining

101. You are in the early stages of a project to manufacture disposable medical devices. You need a number of engineers including ones with specialties in mechanical, environmental, and systems engineering. In the early stages of this project, your resource pool includes a large number of both junior and senior engineers in the various specialty areas. However, as the project progresses—

   a. Fewer systems engineers will be needed
   b. The resource pool can be limited to those people who are knowledgeable about the project
   c. To complete the project on time, you will continue to require access to a large number of engineers in their specialty areas
   d. You will only need junior level engineers as the senior level people can be used early in the project to mentor and train them

102. A number of approaches can be helpful when estimating resource requirements for activities on a project. Assume you are managing a project and you have already prepared your WBS. When you decomposed your WBS, it had work packages. You then prepared an activity list. Now, you are preparing your schedule and determining your resource requirements. You found there were about 30 activities that you could not estimate with a reasonable degree of confidence, so you and your team decided to use which one of the following approaches to help with these activity resource estimates:

   a. Resource breakdown structure
   b. Published estimating data
   c. Alternatives analysis
   d. Bottom-up estimating

103. To practice effective schedule control, your project team must be alert to any issues that may cause problems in the future. To best accomplish effective schedule control, the team should—

   a. Review work performance information
   b. Allow no changes to the schedule
   c. Update the schedule management plan on a continuous basis
   d. Hold status reviews

104. Obviously as a project manager, you will be making decisions throughout the project. A guideline for decision making is—

   a. Active listening skills to ensure all points of view are heard
   b. Focus on the vision of the project
   c. Gather relevant or critical information
   d. Focus on goals to be served

105. You recognize the importance of the project charter, and it has a number of key benefits including—

   a. A direct way for senior management to formally accept and commit to the project
   b. A method to enable the project manager to assign resources to the project
   c. A way to ensure everyone has a foundational understanding of the project
   d. A way to establish internal agreements in the organization to ensure proper delivery under the contract

106. A watch list of low priority risks is documented in the—

   a. Work performance information
   b. Risk register
   c. Fallback plans
   d. Risk response plan

107. You are the project manager for a major logistics installation project and must obtain specific services from local sources external to your project. Your subcontracts administrator has told you to prepare a product or service description, which is referenced in a—

   a. Project statement of work
   b. Contract scope statement
   c. Request for proposal
   d. Contract

108. Assume you are finally in the Close Project phase after almost two years on your project. You are holding a series of meetings, one of which is—

   a. Team performance assessments
   b. User groups
   c. Requirements analysis
   d. Contractor assessments for the qualified seller list

109. You are working on a project to upgrade the existing fiber-optic cables in your province. You have determined that a resource can install 25 meters of cable per hour, so the duration required to install 1,000 meters would be 40 hours. This means you are using—

   a. Productivity efficiency factors
   b. Parametric estimating
   c. Analogous estimating
   d. PERT

110. During the stages of team development, your team is in which stage when there is problem solving and interdependence along with achievement and synergy?

   a. Storming
   b. Forming
   c. Norming
   d. Performing

111. When you are about to terminate a contract, the one place to look for specific procedures for contract closure is in the—

   a. Statement of work in the contract
   b. Terms and conditions in the contract
   c. Product description
   d. Organizational process assets

112. Assume even though you are the seller, you have asked the buyer to terminate your contract for convenience. You and your team believe, based on the work done so far, you will not be able to satisfy the buyer. You added staff members even though it was a fixed-price contract, but still your deliverables are being rejected. You and your team are in shock, but the buyer agrees. Now, the buyer is working to terminate the contract for convenience. As the buyer one should—

   a. Reassign staff
   b. Have the procurement staff handle the termination and the closing activities
   c. Suggest an increase in the allocated time to the seller
   d. Involve stakeholders

113. You decided to implement a team-based reward and recognition for your project, and the team members agreed on the project's ground rules that were established. However, while your sponsor liked the idea, the human resources department states individual performance appraisals must be done. As the project manager, you do so—

   a. At the time the team member completes work on his or her deliverable or activities
   b. When the project is in the closing phase
   c. As specified in the person's Individual Development Plan
   d. In conjunction with the functional manager

114. Your company is embarking on a project to launch a new product delivery service. You are the project manager for this project and have just finished the concept phase. The various outputs/deliverable(s) for this phase include—

   a. Project management plan
   b. Scope of work and requirements
   c. Project charter and stakeholder register
   d. Roles and responsibilities of the project manager

115. As you work in the executing process, and especially in the Direct and Manage Project Work process, the majority of the project's work is being completed. As an output, you know you need to update a number of project documents, one of which is the—

   a. Project baselines
   b. Issue logs
   c. Key performance indicators
   d. Process improvement plan

116. You are managing a project in which your team members all work in the same geographic location and have worked together previously on many projects. Everyone is aware of the various strengths and weaknesses of the individual team members and their key areas of expertise. As a result—

   a. A kickoff meeting is recommended
   b. Team-building activities will not be needed on your project
   c. You should expect minimal conflicts and changes to occur
   d. Rewards and recognition will be handled smoothly throughout the project

117. Team building should be ongoing throughout the project life cycle. However, it is hard to maintain momentum and morale, especially on large, complex projects that span several years. One guideline to follow to promote team building is to—

   a. Consider every meeting a team meeting, not the project manager's meeting
   b. Conduct team building at specific times during the project through off-site meetings
   c. Engage the services of a full-time facilitator before any team-building initiatives are conducted
   d. Develop the project schedule using the services of a project control officer and then issue it immediately to the team

118. You have been assigned as the project manager for a major project in your company where the customer and key supplier are located in another country. You have been working on your project for six months. Recently, you traveled to this country, and at the conclusion of a critical design review meeting, which was highly successful, you realized you were successful in building a high-performing team. You had your own team members, who work in a weak matrix structure, on a conference call during this meeting. Although it was difficult to reach agreement on some key issues, you therefore relied on your interpersonal skills in—

    a. Facilitation
    b. Negotiating
    c. Influencing
    d. Decision making

119. You feel fortunate to be assigned as the project manager on a multi-phase project that was requested by your company's key customer. It has the interest of the senior executives, and it was approved by the organization's portfolio oversight group. As you begin to work on this multi-phase project, a best practice is to—

    a. Periodically review the business case
    b. Establish a Governance Board to conduct the phase reviews
    c. Use project audits
    d. Focus on realizing benefits

120. Because risk management is relatively new on projects in your company, you decide to examine and document the effectiveness of risk responses in dealing with identified risks and their root causes. You therefore—

    a. Conduct a risk audit
    b. Hold a risk status meeting
    c. Ensure that risk is an agenda item at regularly scheduled staff meetings
    d. Reassess identified risks on a periodic basis

121. Thinking back to lessons that your company learned from experiences with its legacy information systems during the Y2K dilemma, you finally convinced management to consider systems maintenance from the beginning of the project. However, maintenance should—

    a. Always be included as an activity to be performed during the closeout phase
    b. Have a separate phase in the life cycle for information systems project because 60% to 70% of computer systems' life-cycle costs generally are devoted to maintenance
    c. Not be viewed as part of the project life cycle
    d. Be viewed as a separate project

122. On your systems development project, you noted during a review that the system had less functionality than planned at the critical design review. This note suggests that during the Control Risks process you used which following tools and techniques?

  a. Risk reassessment
  b. Variance analysis
  c. Technical performance measurement
  d. Reserve analysis

123. The workaround that you used to deal with a risk that occurred should be documented and included in which following processes?

  a. Report Performance
  b. Validate Scope
  c. Direct and Manage Project Work
  d. Control Risks

124. Contested changes are requested changes when the buyer and seller cannot agree on compensation for the change. They are also known as all but which one of the following?

  a. Disputes
  b. Demands
  c. Appeals
  d. Claims

125. A structured review of the seller's progress to deliver project scope and quality within cost and schedule is known as a(n)—

  a. Procurement performance review
  b. Procurement audit
  c. Inspection
  d. Status review meeting

126. Within your company's portfolio, your project is ranked in the top five in terms of importance of the 60 projects under way; however, the number of resources available to you is still limited. You have decided to pilot test the use of critical chain on your project. You have calculated your critical path. You want to ensure that your target finish date does not slip in the critical chain method. To do so you should—

  a. Add a project buffer
  b. Put in three feeding buffers
  c. Determine the drum resource
  d. Manage the total float of the network paths

127. Assume you are developing your project charter and decided to review enterprise environmental factors. An example is—

  a. Codes of conduct
  b. Process definitions
  c. Information from the risk management activity
  d. Results of previous project selection decisions

128. A team-building approach that facilitates a sense of community is—

  a. Matrix management
  b. Fast-tracking
  c. Tight matrix
  d. Task force

129. On your project you decided it would be worthwhile to prepare a stakeholder engagement matrix. You felt this matrix would help show gaps between the current and the desired engagement levels of your project's stakeholders. After you prepared this matrix, you decided to work with your team to determine actions and communications required to close the identified gaps. You did so by—

  a. Updating the stakeholder register
  b. Consulting with experts
  c. Assigning a team member to work with stakeholders in each category
  d. Showing identified interrelationships and potential overlaps among the stakeholders

130. Historical information is used—

  a. To compare current performance with prospective lessons learned
  b. To prepare the stakeholder management plan
  c. To evaluate the skills and competencies of prospective team members
  d. As an input to Develop Project Charter

131. Work completed, key performance indicators, technical performance measures, start and finish dates of schedule activities, number of change requests, number of defects, actual costs, and actual decisions are examples of work performance data output of—

  a. Project Plan Development
  b. Risk Control
  c. Monitor and Control Project Work
  d. Direct and Manage Project Work

132. Two team members on your current construction project are engaged in a major argument concerning the selection of project management software. They refuse to listen to each other. The most appropriate conflict resolution approach for you to use in this situation is—

a. Accommodating
b. Compromising
c. Collaborating
d. Forcing

133. As you use the critical chain method in lieu of the critical path method in developing your schedule, assume you have determined the buffer schedule activities. Your planned activities are scheduled to their latest possible planed start and end dates. Therefore, you are focusing on—

a. Managing the free float of each network path
b. Managing the total float of the network paths
c. Managing remaining buffer durations against the remaining durations of task chains
d. Managing the total buffer durations against the durations of the task chains

134. A key member of your project has deep technical skills and many years of experience in the company. However, she felt she should be the project manager. When you became the project manager instead, her morale deteriorated. You worked with her to obtain her ideas whenever there were issues and commended her work to others. But her morale is so low, and she is constantly complaining. Now you notice her morale is so poor that it is affecting other team members, and there are numerous negative conflicts you need to resolve. You have decided you need to reassign this staff member and have worked with your PMO manager to do so. Your next step is to—

a. Meet with this team member and inform her of your decision
b. Inform the team she is moving to a new position but recognize her contributions to them
c. Issue a change request
d. Work with human resources and then meet with this team member

135. You are performing a stakeholder analysis on your project, and you are working with your team to ensure you do not omit a key stakeholder. Your first step should be to—

a. Prepare a stakeholder register
b. Determine the model you plan to use to classify the stakeholders
c. Assess how certain stakeholders are likely to respond in certain situations
d. Identify all potential project stakeholders and their relevant information

136. Before considering a project closed, what document should be reviewed to ensure that project scope has been satisfied?

    a. Project scope statement
    b. Project management plan
    c. Project closeout checklists
    d. Scope management plan

137. A cost management plan should establish and document the various earned value rules of performance measurement. Along with defining the WBS to the level that the earned value analysis will be performed and establishing how earned value will be credited to the project (0-100, 0-50-100, and so on), which following rule is also recognized regarding performance measurement?

    a. Determine the formula for calculating the estimate to complete (ETC) for the project
    b. Determine the code of accounts allocation provision for the WBS
    c. Determine the formula for calculating the estimate at completion (EAC) for the project
    d. Determine the variance thresholds to be used in the project

138. All the following elements are organizational process asset updates, resulting from closing a project or phase EXCEPT—

    a. Project files
    b. Project or phase closure documents
    c. Historical information
    d. Final product, service, or result transition

139. You have a conflict on your team but have enough time to resolve it, and you want to maintain future relationships. Thankfully, there is mutual trust, respect, and confidence among the parties involved. You decide to use confronting to resolve this conflict. In using this approach, your first step should be to—

    a. Separate people from the problem
    b. Identify the causes of the conflict
    c. Establish ground rules
    d. Explore alternatives

140. One way to evaluate the project schedule performance is to—

    a. Use the project management information system (PMIS)
    b. Determine the percent complete of in-progress schedule activities
    c. Establish a schedule change control system
    d. Determine the total float variance

141. Validate Scope works hand-in-hand with Control Quality and generally follows Control Quality. A tool and technique used in Validate Scope that is not used in Control Quality is—

   a. Group decision-making techniques
   b. Inspection
   c. Statistical sampling
   d. Variance analysis

142. Assume your project is considered to be extremely important to your company as it is for its top client. You have been given the authority to assign resources to the project as you set up your team. An important criterion is—

   a. Experience
   b. Ability
   c. Knowledge
   d. Skills

143. You are a goal-oriented project manager who is more interested in work accomplishment than relationship building. This indicates that you tend to resolve conflicts primarily through the use of—

   a. Smoothing
   b. Compromising
   c. Collaborating
   d. Forcing

144. You are working on a long-term project that has a number of benefits to its customers and users. Therefore, as the project manager, one of your first steps was to identify the stakeholders that were critical to project success. Because this project will need long-term support by your organization once it is completed, key stakeholders are—

   a. Operations managers
   b. Manufacturing managers
   c. Sellers
   d. Business partners

145. Effective communication occurs in groups as well as between individuals and is made up of several key components, such as the purpose of the message, the audience that you are delivering the message to, and the content of the message itself. One important area to consider when working with manage communications—

   a. The choice of media
   b. How often to distribute the information
   c. The communications plan
   d. The project performance report structure

146. A conflict resolution approach that is NOT considered to be very effective when more than a few players are involved and their viewpoints are mutually exclusive is—

   a. Forcing
   b. Avoiding
   c. Compromising
   d. Collaborating

147. The key output of Identify Stakeholders that documents identification information, assessment information, and classification is the—

   a. Stakeholder management plan
   b. Communications plan
   c. Stakeholder register
   d. Communications log

148. Improvement to the processes and the product is a goal of project quality management. Assume that after completing a quality audit, you have discovered some gaps/shortcomings in the way that the project team is completing one deliverable. As an output to Perform Quality Assurance, you would create which of the following item that feeds directly into the Perform Integrated Change Control processes?

   a. Quality management plan updates
   b. Risk register
   c. Change requests
   d. Project document updates

149. The Estimate Costs process uses all the following tools and techniques EXCEPT—

   a. Three-point estimating
   b. Cost of quality assumptions
   c. Reserve analysis (contingency reserves)
   d. Basis of estimates

150. A contract is a type of agreement, typically used when a project is being performed for an external customer. Agreements of all types are used as an input to—

   a. Develop Project Charter
   b. Develop Project Team
   c. Plan Procurement Management
   d. Conduct Procurements

151. As you prepare to close your project, which of the following is an input to the Close Project or Phase process?

 a. Work performance information
 b. Expert judgment
 c. Accepted deliverables
 d. Change requests

152. Managing change to the scope baseline is the main benefit of the process of Control Scope. As you work to avoid scope creep on you project, you want to focus on determining the cause and degree of difference from this baseline and project performance. To do so, you decide to—

 a. Establish scope guidelines
 b. Set up and follow a requirements traceability matrix
 c. Set up scope categories
 d. Use variance analysis

153. Assume you have been working with your sponsor to prepare you charter, and you plan to present it to your Steering Committee on Friday. You are managing a software project, and the business need stated that you should use agile for the first time in your company rather than waterfall. In the Develop Project Charter process, this is then—

 a. A tool and technique
 b. Part of the enterprise environment factors as an input to this process
 c. A high-level requirement
 d. Stated in the strategic plan as a tool and technique in this process

154. Procurement documents are used in the Identify Stakeholder process because they—

 a. Are an enterprise environmental factor and an input to the process
 b. Are an organizational process asset and an input to the process
 c. Note key stakeholders as parties in the contract
 d. Serve as a way to prioritize and classify stakeholders

155. You completed your stakeholder analysis. How do you want to manage those stakeholders that have a high interest in your project and high power over decisions affecting your project?

 a. Manage them closely
 b. Keep them satisfied
 c. Keep them informed
 d. Monitor them occasionally

156. Change requests include a group of potential changes to a project. Types of change requests include all the following EXCEPT—

    a. Updates
    b. Maintenance requests
    c. Corrective actions
    d. Preventive actions

157. You are working on a project that needs approval from your City Council and the courts, because the project is one with significant environmental and social impacts. Although many consumer groups are advocates of this project, others are opposed to it. Hearings are scheduled to resolve these issues and to obtain the needed permits to proceed. In preparing your human resource plan, you decide to designate a person as the court liaison, which is an example of a—

    a. Role
    b. Responsibility
    c. Required competency
    d. Ability of the team member to make appropriate decisions

158. Assume you are managing an international project. Your team is located in Atlanta, Georgia, US; Berlin, Germany; and Melbourne, Australia. You and your sponsor are located in Paris, France, and your customer is located in Athens, Greece. Recognizing the different locations of the stakeholders in your project in its initial stages, a best practice to follow in terms of working toward project success is to—

    a. Determine who decides the project is a success
    b. Aligning the personal inputs of different project participants with a vision focused on success
    c. Establishing the project culture during the initiating stage of the project
    d. Identifying basic cultural characteristics and selecting one to follow

159. As a project manager, you recognize the importance of actively engaging key project stakeholders on a project. You have prepared an analysis of your stakeholders early in your project and classified them according to their interest, influence, and involvement in your project. You want to now—

    a. Focus on relationships necessary to ensure success
    b. Assess stakeholder legitimacy
    c. Determine the urgency that each stakeholder requires when he or she requests information about the project
    d. Focus on each stakeholder's power relevant to the project

160. As you prepare your human resource management plan for your project, you have decided to document the roles and responsibilities of your eight team members. You have looked at some formats to use and find most are in the form of an organization chart, the responsibility chart or matrix, or a role description in text format. However, regardless of the format, the emphasis is to—

    a. List resources based on category
    b. Track project costs and align them with the organization's accounting system
    c. Have an unambiguous owner for each work package
    d. Enable an operational department to easily see all of its portion of the project's resources

161. The Plan Quality Management process includes all the following techniques EXCEPT—

    a. Benchmarking
    b. Design of experiments
    c. Process analysis
    d. Control charts

162. You are managing a major international project that involves multiple performing organizations. To establish the guiding rules for the project regarding quality, you and your project team must develop a—

    a. Improvement management plan
    b. Configuration management plan
    c. Quality policy
    d. List of quality metrics for the project

163. You are working on a project that management has decided to terminate early, because the product was rendered obsolete by the introduction of new technology by a competitor. You have awarded a contract for part of the project that will be terminated, and fortunately have a clause that enables you to terminate it for convenience at any time. This means that—

    a. Your contractual obligations are complete once you issue the termination for convenience
    b. You may need to compensate the seller for seller preparations and for any completed or accepted work
    c. You need to compensate the seller only for accepted work that was completed prior to the termination order
    d. Specific rights and responsibilities are determined once the termination order is issued

164. Of the following, which one is NOT true concerning an agreement?

   a. It is a legal relationship subject to remedy in the courts.
   b. It can take the form of a complex document or a simple purchase order.
   c. It is a mutually binding legal relationship that obligates the seller to provide specific products, services, or results and obligates the buyer to pay the seller.
   d. It includes a specific contract management plan.

165. All of the following can be used in lieu of the term "bidders conferences" EXCEPT—

   a. Contractor conferences
   b. Pre-bid conferences
   c. Vendor conferences
   d. Project review meetings

166. Assume you are preparing your procurement management plan. A useful tool and technique is to—

   a. Have a meeting
   b. Use your risk register
   c. Review the requirements document
   d. Use your stakeholder register

167. Your role in the project includes helping to resolve problems; making recommendations regarding priorities; accelerating activities to meet the target schedule; promoting communications among project team members; and helping management monitor the project's progress on a regular basis. Most of the people working on your project are scientists or technical experts. You are working in which of the following types of organizational structures?

   a. Task force
   b. Balanced matrix
   c. Project expeditor
   d. Project coordinator

168. Assume you are working on a complex project in your organization, and it involves in-house staff, as well as contractors. Some of your team members work in different locations. At first, the team members did not really understand their roles and responsibilities and were not that committed to the project. To avoid cost and schedule overruns, you learned you needed to have a lot of meetings with your team to obtain their buy-in and use more of a command and control leadership approach. Now, however, a year has passed, and the project is on schedule and within budget. There is greater trust among the team members. You have learned since this team is more mature that you—

   a. Can use an adaptive project life cycle
   b. Should focus on networking and professional development
   c. Can use a flexible leadership style
   d. Can promote interface relationship management

169. While you have worked as a project manager for ten years, you are new to your current organization. In the past, your organization was informal in terms of plans to prepare and processes to follow. Your new organization, however, requires that each project have a human resource management plan. You are unsure of the reporting relationships that need to be followed. You decide to—

   a. Hold meetings with the team members who have been assigned thus far
   b. Use experts
   c. Develop a RACI chart and review it with your sponsor
   d. Use your WBS and develop a hierarchical organization chart

170. An intentional activity to ensure future performance of project work is aligned with the project management plan is—

   a. Preventive action
   b. Corrective action
   c. Implemented change requests
   d. Work performance information

171. Close procurements is a process that involves includes activities for administrative closure such as—

   a. Customer acceptance and final payment
   b. Auditing project success or failure and archiving records
   c. Final contractor payment and lessons learned
   d. Transition of the final product and acceptance of deliverables

172. As a project manager, not only must you be a leader, but you also must have outstanding skills in communicating because research over the years shows project managers spend about 90% of their time communicating. You are working to create, collect, distribute, store, retrieve, and ultimately dispose of project information as described in your communications management plan. The benefit of this work is—

    a. An effective and efficient flow of communications among stakeholders
    b. Creating an environment with an emphasis on transparent communications with your stakeholders
    c. Having resources available to ensure your stakeholders have access to needed information in a timely way
    d. Ensuring communications management policies, operating procedures, and processes are followed

173. Assume you are managing a project, and your project management plan has been approved. Your project has a high level of change associated with it. There is active and ongoing stakeholder involvement. This means you probably are working with a(n)—

    a. Adaptive life cycle
    b. Iterative life cycle
    c. Incremental life cycle
    d. Predictive life cycle

174. One way to help mitigate personnel risks that may occur during the project is to—

    a. Meet individually with each team member
    b. Provide specific recognition to each team member who has worked on the project
    c. Prepare a staff release plan
    d. Document the time each person is to work on the project in a resource calendar

175. You are conducting a stakeholder analysis on your project. After identifying potential stakeholders, the next step in the process is to—

    a. Determine their desired level of participation
    b. Provide detailed contact information for each identified stakeholder
    c. Perform an assessment to see how each stakeholder might react in certain situations
    d. Analyze each stakeholder's impact or support and classify them

176. Assume your project communication management plan has been approved by your sponsor and the members of your Steering Committee. You are managing a global project and have team members working virtually in four continents and stakeholders in numerous locations. Your next step is to—

    a. Set up an information management system
    b. Select communications technology
    c. Determine performance reporting methods
    d. Select a communications model

177. Although your project team is working virtually, you are striving to make it a high-performing team. You held a virtual kickoff meeting to ensure there was a shared project vision. You now see that team members are addressing the work to be done, but they do not seem to be collaborating. You realize the team is—

    a. Concerned about their formal roles and responsibilities
    b. Independent
    c. Forming
    d. Storming

178. You are conducting a stakeholder analysis on your project. Your organization uses an approach to classify stakeholders based on their level of authority and their active involvement in the project. This approach is known as—

    a. A power/interest grid
    b. A power/influence grid
    c. An influence/impact grid
    d. A salience model

179. When managing projects, one important technique is capturing lessons learned from previous projects to improve the organization's project management process. Therefore, assume you are working to identify possible risks to your project, and there are numerous techniques to use but you should consider—

    a. Expert judgment
    b. Checklists
    c. Influence diagrams
    d. Interviews

180. Failure to meet quality requirements can have serious, negative effect for a project and its stakeholders. Therefore, you are focusing on quality assurance as an ongoing activity in your projects. It helps ensure appropriate standards and operational definitions are used, which means—

   a. A focus on Six-Sigma, failure mode and effect analysis, and total quality management
   b. An emphasis on the International Organization for Standardization
   c. Using methods, such as those recommended by Deming, Juran, and Crosby
   d. Focusing on inspection over prevention

181. During a bidders conference, it is important that—

   a. Only qualified sellers participate
   b. All potential sellers are given equal standing
   c. The evaluation criteria for the proposal is used to determine participation
   d. Responses to questions be provided solely to the prospective seller that asked the question

182. One key interpersonal skill used to Manage Stakeholder Expectations is—

   a. Negotiation skills
   b. Building trust
   c. Compromise
   d. Conversation

183. You are a project manager leading the construction project of a new garbage incinerator. Local residents and environmental groups are opposed to this project because of its environmental impact. Management agrees with your request to partner with a third party that will be responsible for providing state-of-the-art "air scrubbers," to clean the exhaust to an acceptable level. This decision will delay the project but will allow it to continue. It is an example of which of the following risk response?

   a. Passive acceptance
   b. Active acceptance
   c. Mitigation
   d. Transference

184. After a year and a half, it is now time for you to close your program. First, you are working on closing procurements. The key benefit of the Close Procurement process is—

   a. It documents agreements and related documentation for future reference
   b. It ensures contractual agreements are completed or terminated
   c. It finalizes any open claims
   d. It ensures there is an equitable settlement of any disputes

185. Working in the systems integration field, you are primarily responsible for coordinating the work of numerous subcontractors. Your current project is coming to an end. You have 15 major subcontractors as well as a variety of other sellers. Now that you are closing contracts (procurements), you should—

a. Conduct a trend analysis
b. Use earned value to assess lessons learned
c. Ask each contractor to meet with you individually at its own expense
d. Conduct a procurement audit

186. You are managing a moderately risky project. You have done well identifying risks and assessing them both on the probability of the risk event occurring and on the level of impact that the risk could have on your project if it actually occurred. But to help with a "next" level of qualitative risk analysis, you could use a—

a. Risk priority assessment
b. Risk quality assessment
c. Risk urgency assessment
d. Quantitative risk analysis

187. You are identifying possible risks to your project concerning the development of a nutritional supplement. You want to reach out to your experts and build consensus on the risks that qualitative and quantitative risk analysis can address later. Although you can use various techniques, a key information gathering technique that helps to build consensus is—

a. Documentation review
b. Probability/impact analysis
c. Checklist analysis
d. Delphi technique

188. Managing five contractors on your project for a new stadium in your City that can be used for baseball and for football and can be easily converted for either sport is a challenge along with managing your 15 person project team. You decided to conduct an audit of one of your contractors and are—

a. Using it as a baseline for improvements to the other contracts under way
b. Verifying compliance in the seller's work processes
c. Accompanying it with a project quality audit
d. Using it for lessons learned documentation

189. You are awarding another contract to serve as an integration contract on your stadium project. It has generated a lot of interest from potential sellers, and you now expect a number of proposals. You decide a weighting system can be used for all but which one of the following reasons?

    a. To select a single seller that will be asked to sign a standard contract
    b. To establish a negotiating sequence by ranking all proposals by the weighted evaluation scores that have been assigned
    c. To establish a short list of qualified sellers
    d. To establish minimum requirements of performance for one or more of the evaluation criteria

190. One way to make it more likely to practice project risk management on projects is to—

    a. Hold meetings
    b. Have the team take an orientation class on risk management early in the project
    c. Give one team member the responsibility for risk management working in conjunction with the PMO
    d. Assign each team member a risk to own, which is documented in the risk register

191. Your firm specializes in roller-coaster construction. It recently received an RFP to build the world's most "death-defying" roller coaster. You know that such a roller coaster has never been built before and that this would be a high-risk project. If your firm wins, it will receive a cost-plus-award-fee contract, which means—

    a. Your fee will be paid for completed work
    b. Your fee amount will not change unless there is a scope change
    c. Your fee is generally not subject to appeals
    d. If your final cost is greater, you will share costs with the buyer based on a pre-negotiated cost sharing formula

192. Work performance information in Control Risks means that—

    a. Recommended preventive or corrective action is considered through change requests
    b. Outcomes of risk reassessments and risk audits are documented
    c. Templates to the risk management plan and the risk register are recommended
    d. A mechanism to communicate and support project decision making is provided

193. A number of factors affect make-or-buy decisions such as—

   a. Desired level of quality
   b. Risk-related contract decisions
   c. Value delivered by vendors meeting the needs
   d. Performance data

194. For complex procurement items, often contract negotiation can be an independent process. An example of an input if such a process is used is—

   a. Open items list
   b. Approved changes
   c. Documented decisions
   d. Expert judgment

195. Marketplace conditions are an input to which one of the following processes?

   a. Plan Procurement Management
   b. Conduct Procurements
   c. Control Procurements
   d. Close Procurements

196. Each project can benefit from stakeholder involvement; however, it is in both the project manager's and the teams' best interest to ensure that all project stakeholders have positive attitudes toward the project and its goals and objectives. Working as a project manager, you have a number of key stakeholders on your project. The stakeholder that identifies potential conflicts between organizational strategies and project goals is the—

   a. Chairperson of the Governance Board
   b. Program manager
   c. Director of the project management office
   d. Chief Operating Officer

197. Assume you are awarding a contract for your project for a new generation of nuclear missiles. The contract will have an extensive approval process because—

   a. It is a legally binding agreement
   b. Services of experts will be needed
   c. The contract language describes what is necessary to satisfy the project's needs
   d. A number of stakeholders must sign off before the award is made

198. When determining the message that you will deliver to stakeholders, knowing both the content (what you want to say) and your audience is important. Which of the following helps you to understand how others may interpret your message?

 a. Sender-receiver models
 b. Facilitation techniques used in delivery
 c. Negotiation skills
 d. Presentation skills used in the development of the message

199. One of the reasons why it is challenging to work on a virtual team is that e-mail is the primary form of communications. However, words alone typically comprise what percent of the total impact of any message?

 a. Seven percent
 b. 15 percent
 c. 38 percent
 d. 55 percent

200. Assume you are working for a major airline in your country. Rather than use paper tickets to board, it has authorized a new product so they are available on mobile phones or smart watches to show to the gate agent and speed up boarding times. This project then is authorized because of a—

 a. Customer request
 b. Organizational need
 c. Market demand
 d. Technological advance

Answer Sheet

1.	a	b	c	d	21.	a	b	c	d
2.	a	b	c	d	22.	a	b	c	d
3.	a	b	c	d	23.	a	b	c	d
4.	a	b	c	d	24.	a	b	c	d
5.	a	b	c	d	25.	a	b	c	d
6.	a	b	c	d	26.	a	b	c	d
7.	a	b	c	d	27.	a	b	c	d
8.	a	b	c	d	28.	a	b	c	d
9.	a	b	c	d	29.	a	b	c	d
10.	a	b	c	d	30.	a	b	c	d
11.	a	b	c	d	31.	a	b	c	d
12.	a	b	c	d	32.	a	b	c	d
13.	a	b	c	d	33.	a	b	c	d
14.	a	b	c	d	34.	a	b	c	d
15.	a	b	c	d	35.	a	b	c	d
16.	a	b	c	d	36.	a	b	c	d
17.	a	b	c	d	37.	a	b	c	d
18.	a	b	c	d	38.	a	b	c	d
19.	a	b	c	d	39.	a	b	c	d
20.	a	b	c	d	40.	a	b	c	d

41.	a	b	c	d
42.	a	b	c	d
43.	a	b	c	d
44.	a	b	c	d
45.	a	b	c	d
46.	a	b	c	d
47.	a	b	c	d
48.	a	b	c	d
49.	a	b	c	d
50.	a	b	c	d
51.	a	b	c	d
52.	a	b	c	d
53.	a	b	c	d
54.	a	b	c	d
55.	a	b	c	d
56.	a	b	c	d
57.	a	b	c	d
58.	a	b	c	d
59.	a	b	c	d
60.	a	b	c	d

61.	a	b	c	d
62.	a	b	c	d
63.	a	b	c	d
64.	a	b	c	d
65.	a	b	c	d
66.	a	b	c	d
67.	a	b	c	d
68.	a	b	c	d
69.	a	b	c	d
70.	a	b	c	d
71.	a	b	c	d
72.	a	b	c	d
73.	a	b	c	d
74.	a	b	c	d
75.	a	b	c	d
76.	a	b	c	d
77.	a	b	c	d
78.	a	b	c	d
79.	a	b	c	d
80.	a	b	c	d

81.	a	b	c	d	101.	a	b	c	d
82.	a	b	c	d	102.	a	b	c	d
83.	a	b	c	d	103.	a	b	c	d
84.	a	b	c	d	104.	a	b	c	d
85.	a	b	c	d	105.	a	b	c	d
86.	a	b	c	d	106.	a	b	c	d
87.	a	b	c	d	107.	a	b	c	d
88.	a	b	c	d	108.	a	b	c	d
89.	a	b	c	d	109.	a	b	c	d
90.	a	b	c	d	110.	a	b	c	d
91.	a	b	c	d	111.	a	b	c	d
92.	a	b	c	d	112.	a	b	c	d
93.	a	b	c	d	113.	a	b	c	d
94.	a	b	c	d	114.	a	b	c	d
95.	a	b	c	d	115.	a	b	c	d
96.	a	b	c	d	116.	a	b	c	d
97.	a	b	c	d	117.	a	b	c	d
98.	a	b	c	d	118.	a	b	c	d
99.	a	b	c	d	119.	a	b	c	d
100.	a	b	c	d	120.	a	b	c	d

121.	a	b	c	d
122.	a	b	c	d
123.	a	b	c	d
124.	a	b	c	d
125.	a	b	c	d
126.	a	b	c	d
127.	a	b	c	d
128.	a	b	c	d
129.	a	b	c	d
130.	a	b	c	d
131.	a	b	c	d
132.	a	b	c	d
133.	a	b	c	d
134.	a	b	c	d
135.	a	b	c	d
136.	a	b	c	d
137.	a	b	c	d
138.	a	b	c	d
139.	a	b	c	d
140.	a	b	c	d

141.	a	b	c	d
142.	a	b	c	d
143.	a	b	c	d
144.	a	b	c	d
145.	a	b	c	d
146.	a	b	c	d
147.	a	b	c	d
148.	a	b	c	d
149.	a	b	c	d
150.	a	b	c	d
151.	a	b	c	d
152.	a	b	c	d
153.	a	b	c	d
154.	a	b	c	d
155.	a	b	c	d
156.	a	b	c	d
157.	a	b	c	d
158.	a	b	c	d
159.	a	b	c	d
160.	a	b	c	d

161.	a	b	c	d	181.	a	b	c	d
162.	a	b	c	d	182.	a	b	c	d
163.	a	b	c	d	183.	a	b	c	d
164.	a	b	c	d	184.	a	b	c	d
165.	a	b	c	d	185.	a	b	c	d
166.	a	b	c	d	186.	a	b	c	d
167.	a	b	c	d	187.	a	b	c	d
168.	a	b	c	d	188.	a	b	c	d
169.	a	b	c	d	189.	a	b	c	d
170.	a	b	c	d	190.	a	b	c	d
171.	a	b	c	d	191.	a	b	c	d
172.	a	b	c	d	192.	a	b	c	d
173.	a	b	c	d	193.	a	b	c	d
174.	a	b	c	d	194.	a	b	c	d
175.	a	b	c	d	195.	a	b	c	d
176.	a	b	c	d	196.	a	b	c	d
177.	a	b	c	d	197.	a	b	c	d
178.	a	b	c	d	198.	a	b	c	d
179.	a	b	c	d	199.	a	b	c	d
180.	a	b	c	d	200.	a	b	c	d

Appendix: Study Matrix

Overview

Periodically, the Project Management Institute (PMI®) publishes a Role Delineation Study (RDS) and uses to define the responsibilities of the recipients of the Project Management Professional (PMP®) credential, In June 2015, PMI issued a new RDS, for the PMP®, which then was published as the PMI® PMP® *Examination Content Outline* (ECO). You can download it from PMI's web site under the Certification section and then by the PMP©. It serves as the foundation for the PMP® exam and for our 200-question practice test in this book and in our online test. This current RDS is in effect for the PMP© exam beginning November 1, 2015.

The ECO identified five broad performance domains and determined how the 175 questions on the PMP® exam would be distributed according to these domains*. The distribution is as follows:

I	Initiating	13%
II	Planning	24%
III	Executing	31%
IV	Monitoring and Controlling	25%
V	Closing	7%

The matrix beginning on page identifies each practice test question according to its performance domain and its knowledge area in the *PMBOK® Guide.*

The matrix is designed to help you—

- Assess your strengths and weaknesses in each of the performance domains
- Identify those areas in which you need additional study before you take the PMP® exam

* PMI® distributes 25 pretest questions across the five domains in any way that it deems appropriate for the purpose of "testing" the questions.

Here is an easy way to use the matrix:

Step 1 Circle all the questions you missed on the practice test in Column 1.
Step 2 For each circled question, note the corresponding process in Column 2.
Step 3 To determine whether any patterns emerge indicating weak areas, tally the information you obtained from the matrix.
Step 4 To ensure that you have a good understanding of the major management processes that define a particular knowledge area, including the inputs, tools and techniques, and outputs, refer to the appropriate knowledge area in the *PMBOK® Guide.*

The last column in the matrix is provided for your notes.

Study Matrix

Practice Test Question Number	*Performance Domain Process*	*Knowledge Area*	*Study Notes*
1	Planning	Integration	
2	Executing	Stakeholders	
3	Executing	Procurement	
4	Planning	Scope	
5	Planning	Integration	
6	Monitoring and Controlling	Procurement	
7	Executing	Integration	
8	Planning	Integration	
9	Planning	Scope	
10	Executing	Quality	
11	Executing	Quality	
12	Monitoring and Controlling	Procurement	
13	Monitoring and Controlling	Integration	
14	Planning	Procurement	
15	Monitoring and Controlling	Integration	
16	Initiating	Integration	
17	Executing	Quality	
18	Monitoring and Controlling	Integration	
19	Monitoring and Controlling	Integration	
20	Monitoring and Controlling	Integration	
21	Planning	Scope	
22	Executing	Integration	
23	Monitoring and Controlling	Integration	
24	Executing	Quality	
25	Monitoring and Controlling	Scope	
26	Closing	Integration	
27	Executing	Quality	
28	Planning	Risk	
29	Planning	Scope	

Practice Test Question Number	*Performance Domain Process*	*Knowledge Area*	*Study Notes*
30	Executing	Quality	
31	Monitoring and Controlling	Integration	
32	Initiating	Integration	
33	Executing	Human Resources	
34	Monitoring and Controlling	Time	
35	Monitoring and Controlling	Cost	
36	Monitoring and Controlling	Cost	
37	Planning	Scope	
38	Closing	Integration	
39	Planning	Scope	
40	Monitoring and Controlling	Cost	
41	Executing	Human Resources	
42	Monitoring and Controlling	Cost	
43	Monitoring and Controlling	Cost	
44	Initiating	Integration	
45	Planning	Integration	
46	Executing	Integration	
47	Monitoring and Controlling	Cost	
48	Planning	Scope	
49	Executing	Communications	
50	Monitoring and Controlling	Cost	
51	Executing	Human Resources	
52	Planning	Scope	
53	Monitoring and Controlling	Time	
54	Planning	Risk	
55	Executing	Human Resources	
56	Monitoring and Controlling	Cost	
57	Monitoring and Controlling	Cost	
58	Initiating	Integration	
59	Closing	Integration	
60	Planning	Risk	

Practice Test Question Number	*Performance Domain Process*	*Knowledge Area*	*Study Notes*
61	Executing	Human Resources	
62	Planning	Time	
63	Monitoring and Controlling	Cost	
64	Planning	Time	
65	Executing	Stakeholders	
66	Planning	Time	
67	Monitoring and Controlling	Stakeholders	
68	Monitoring and Controlling	Integration	
69	Planning	Scope	
70	Planning	Time	
71	Planning	Time	
72	Planning	Time	
73	Monitoring and Controlling	Time	
74	Executing	Integration	
75	Executing	Human Resources	
76	Closing	Procurement	
77	Monitoring and Controlling	Procurement	
78	Initiating	Integration	
79	Monitoring and Controlling	Quality	
80	Executing	Human Resources	
81	Monitoring and Controlling	Quality	
82	Executing	Human Resources	
83	Monitoring and Controlling	Quality	
84	Monitoring and Controlling	Quality	
85	Executing	Human Resources	
86	Monitoring and Controlling	Time	
87	Executing	Integration	
88	Initiating	Integration	
89	Executing	Communications	
90	Monitoring and Controlling	Quality	
91	Monitoring and Controlling	Quality	

Practice Test Question Number	*Performance Domain Process*	*Knowledge Area*	*Study Notes*
92	Planning	Cost	
93	Planning	Cost	
94	Executing	Human Resources	
95	Closing	Procurement	
96	Monitoring and Controlling	Quality	
97	Planning	Time	
98	Executing	Stakeholders	
99	Monitoring and Controlling	Stakeholders	
100	Monitoring and Controlling	Time	
101	Planning	Time	
102	Planning	Time	
103	Monitoring and Controlling	Time	
104	Executing	Human Resources	
105	Initiating	Integration	
106	Monitoring and Controlling	Risk	
107	Initiating	Integration	
108	Closing	Integration	
109	Planning	Time	
110	Executing	Human Resources	
111	Closing	Procurement	
112	Closing	Integration	
113	Closing	Integration	
114	Initiating	Integration	
115	Executing	Human Resources	
116	Planning	Human Resources	
117	Executing	Human Resources	
118	Monitoring and Controlling	Human Resources	
119	Initiating	Integration	
120	Monitoring and Controlling	Risk	
121	Initiating	Integration	
122	Monitoring and Controlling	Risk	

Practice Test Question Number	*Performance Domain Process*	*Knowledge Area*	*Study Notes*
123	Monitoring and Controlling	Risk	
124	Monitoring and Controlling	Procurement	
125	Monitoring and Controlling	Procurement	
126	Planning	Time	
127	Initiating	Integration	
128	Executing	Human Resources	
129	Planning	Stakeholders	
130	Initiating	Integration	
131	Executing	Integration	
132	Executing	Human Resources	
133	Planning	Time	
134	Executing	Human Resources	
135	Initiating	Stakeholders	
136	Closing	Integration	
137	Planning	Cost	
138	Closing	Integration	
139	Executing	Human Resources	
140	Monitoring and Controlling	Time	
141	Monitoring and Controlling	Scope	
142	Executing	Human Resources	
143	Executing	Human Resources	
144	Initiating	Stakeholders	
145	Executing	Communications	
146	Executing	Human Resources	
147	Initiating	Stakeholders	
148	Executing	Quality	
149	Planning	Cost	
150	Initiating	Integration	
151	Closing	Integration	
152	Monitoring and Controlling	Scope	
153	Initiating	Integration	

Practice Test Question Number	*Performance Domain Process*	*Knowledge Area*	*Study Notes*
154	Initiating	Stakeholders	
155	Initiating	Stakeholders	
156	Executing	Integration	
157	Planning	Human Resources	
158	Initiating	Integration	
159	Initiating	Stakeholders	
160	Human Resources	Quality	
161	Planning	Quality	
162	Planning	Quality	
163	Executing	Procurement	
164	Executing	Procurement	
165	Executing	Procurement	
166	Planning	Procurement	
167	Planning	Human Resources	
168	Planning	Human Resources	
169	Planning	Human Resources	
170	Executing	Integration	
171	Closing	Procurement	
172	Executing	Communications	
173	Executing	Integration	
174	Planning	Human Resources	
175	Initiating	Stakeholders	
176	Executing	Communications	
177	Executing	Human Resources	
178	Initiating	Stakeholders	
179	Planning	Risk	
180	Executing	Quality	
181	Executing	Procurement	
182	Executing	Stakeholders	
183	Planning	Risk	
184	Closing	Procurement	

Practice Test Question Number	*Performance Domain Process*	*Knowledge Area*	*Study Notes*
185	Closing	Procurement	
186	Planning	Risk	
187	Planning	Risk	
188	Monitoring and Controlling	Procurement	
189	Executing	Procurement	
190	Monitoring and Controlling	Risk	
191	Planning	Procurement	
192	Monitoring and Controlling	Risk	
193	Executing	Procurement	
194	Executing	Procurement	
195	Planning	Procurement	
196	Initiating	Stakeholders	
197	Executing	Procurement	
198	Executing	Communications	
199	Executing	Communications	
200	Initiating	Integration	

Answer Key

1. c. It is not a problem at this time. The previous project manager was using the rolling wave planning technique, so you are able to continue defining the activities.

   Rolling wave planning provides progressive detailing of the work to be accomplished throughout the life of the project. Decomposition is a tool and technique in Create WBS and recognizes rolling wave planning may be needed for deliverables or subcomponents that will not be accomplished until much later in the project. In this case, the team waits until more information about the deliverable or subcomponent is available. [Planning]

   PMI®, *PMBOK® Guide*, 2013, 131
   PMI® *PMP Examination Content Outline*, 2015, Planning, 6, Task 1

2. b. These concerns may become issues

   During the Manage Stakeholder Engagement process, activities include addressing potential stakeholder concerns that have not yet become issues. It involves anticipating future problems that may be raised by stakeholders. It is important, then, to identify and discuss these concerns as soon as possible to assess project risks. [Executing]

   PMI®, *PMBOK® Guide*, 2013, 405
   PMI® *PMP Examination Content Outline*, 2015, Executing, 8, Task 7

3. c. Analytical techniques

   Analytical techniques are a tool and technique in Conduct Procurements. They are used to help organizations identify the readiness of a vendor to provide the desired end state, determine costs to support budgeting, and avoid cost overruns because of changes. By evaluating past performance they identify areas that have more risk and that may need to be monitored closely for project success. [Executing]

   PMI®, *PMBOK® Guide*, 2013, 376
   PMI® *PMP Examination Content Outline*, 2015, Executing, 8, Task 1

4. a. Action requirements

   Such classification systems are helpful in both defining and documenting stakeholder needs to meet project objectives. Project requirements are ones that involve actions, processes, or other conditions the project needs to meet. Solution requirements involve functional and nonfunctional requirements. The functional requirements describe the behaviors of the product. The nonfunctional requirements supplement the functional ones and describe the conditions or qualities required for the product to be effective. The other answers are examples of nonfunctional requirements. [Planning]

   PMI®, *PMBOK® Guide*, 2013, 112
   PMI® *PMP Examination Content Outline,* 2015, Planning, 6, Task 2

5. b. Organizational process assets

   Organizational process assets include formal and informal plans, policies, procedures, and guidelines. As an input to the Develop Project Management plan process, they include the items listed as well as standardized guidelines, instructions, proposal evaluation criteria, and performance measurement criteria; project management plan template; project files from previous projects; and historical information and lessons learned [Planning]

   PMI®, *PMBOK® Guide*, 2013, 75
   PMI® *PMP Examination Content Outline,* 2015, Planning, 6, Task 1

6. a. Control Procurements

   The purpose of Control Procurements is to ensure that the contractual requirements are met by the seller. This objective is accomplished by managing procurement relationships, monitoring contract performance and making changes and corrections to contracts if appropriate. [Monitoring and Controlling]

   PMI®, *PMBOK® Guide*, 2013, 379
   PMI® *PMP Examination Content Outline,* 2015, Monitoring and Controlling, 9, Task 7

7. b. The lowest level of detail to derive information

   Work performance data are an output of the Direct and Manage Project Work process. They are the raw observations and measurements identified during activities being performed to do the project work. Therefore they are often viewed as the lowest level of detail from which to derive information by other processes. The data are gathered and then passed to the controlling processes of each process area for further analysis. [Executing]

   PMI®, *PMBOK® Guide*, 2013, 85
   PMI® *PMP Examination Content Outline,* 2015, Executing, 8, Task 2

8. b. Requirements baseline

   The scope, schedule, and cost baselines may be combined into a performance measurement baseline. It also may include technical and quality parameters. It then is used as an overall project baseline against which project execution is compared, and deviations are managed for project control. It also is used for earned value measurements. [Planning]

   PMI®, *PMBOK® Guide*, 2013, 302, 549
   PMI® *PMP Examination Content Outline,* 2015, Planning, 6, Task 1

9. c. Job shadowing

   Observations are a tool and technique in the Collect Requirements process. They provide a way to view individuals in their environment and to see how they perform their jobs or tasks and carry out processes. Job shadowing usually is done by an observer viewing the user performing his or her job. It can also be done by a 'participant observer' who is performing a process or procedure to experience how it is done to uncover hidden requirements. [Planning]

   PMI®, *PMBOK® Guide*, 2013, 116
   PMI® *PMP Examination Content Outline,* 2015, Planning, 6, Task 2

10. b. Improvement of quality processes

    Perform Quality Assurance involves auditing the quality requirements and the results from quality control measurements to ensure appropriate quality standards and operational definitions are used. The key benefit of this process is it facilitates the improvement of quality processes. It seeks to build confidence that a future output or an unfinished output will be completed in a way that meets specified requirements and expectations. [Executing]

    PMI®, *PMBOK® Guide*, 2013, 242–243
    PMI® *PMP Examination Content Outline,* 2015, Executing, 8, Task 3

11. c. Use an affinity diagram

In quality assurance an affinity diagram is used to generate ideas that can be linked to form organized patterns of thought about a problem. Using them in project management, one can enhance the creation of the WBS by using it to give structure to the decomposition of scope. The affinity diagram is similar to a mind mapping technique. They are part of quality management and control tools, a tool and technique in Perform Quality Assurance. [Executing]

PMI®, *PMBOK® Guide*, 2013, 245
PMI® *PMP Examination Content Outline,* 2015, Executing, 8, Task 3

12. d. Managing the interfaces among the sellers

The Control Procurements process involves managing procurement relationships, monitoring contract performance, and making changes and corrections to contracts as needed. On a large project with multiple sellers, as in this question, a major aspect is managing the interfaces among the sellers. Many organizations have contract administration as an administrative function separate from the project organization. While a procurement administrator may be on the core team, typically this person reports to someone other than the project manager in a different department. [Monitoring and Controlling]

PMI®, *PMBOK® Guide*, 2013, 379–380
PMI® *PMP Examination Content Outline,* 2015, Monitoring and Controlling; 9, Task 7

13. c. Document the specific responsibilities of each stakeholder in the Perform Integrated Change Control process

Configuration management is an integral part of the Perform Integrated Change Control process. It is necessary because projects by their nature involve changes. Configuration control is focused on the specification of deliverables and processes. It emphasizes configuration identification (answer a), configuration status accounting (answer b), and configuration verification and audit (answer d). For further Information refer to PMI's *Practice Standard for Configuration Management,* 2007. [Monitoring and Controlling]

PMI®, *PMBOK® Guide*, 2013, 96–97
PMI® *PMP Examination Content Outline,* 2015, Monitoring and Controlling, 9, Task 2

14. b. Procurement management plan

    The procurement management plan describes how the project management team will acquire goods and services from outside the performing organization. It describes how the procurement processes will be used from developing procurement documents through closing contracts. [Planning]

    PMI®, *PMBOK® Guide,* 2013, 366–367
    PMI® PMP Examination Content Outline, 2015, Planning, 6, Task 7

15. a. Scope, quality, schedule, budget, and risk

    The constraints include, but are not limited to scope, schedule, budget (cost), quality, resources, and risk. Constraints are competing, and specific project characteristics and circumstances can influence them. They are limiting factors that affect project execution so must be monitored and controlled. [Monitoring and Controlling]

    PMI®, *PMBOK® Guide,* 2013, 6, 124
    PMI® *PMP Examination Content Outline,* 2015, Monitoring and Controlling, 9, Task 1

16. d. Ecological impact

    The business case is created in this question primarily as a result of an ecological need. Answer b refers to a customer request such as an electrical utility authorizing a project to build a new substation to serve a new industrial park. Answer c relates to a social need such as a non-governmental organization in a developing country to provide portable water systems plus other items to communities suffering from high rates of cholera. Other reasons are a market demand, an organizational need, a technological advance or a legal requirement. Sustainability does not apply. The business case is an input to the Develop Project Charter process. [Initiating]

    PMI®, *PMBOK® Guide,* 2013, 69
    PMI® *PMP Examination Content Outline,* 2015, Initiating, 5, Task 1

17. d. A quality audit

A quality audit is a tool and technique in the Perform Quality Assurance process. It is primarily used to determine whether the project team is complying with organizational and project policies, processes, and procedures. It identifies good and best practices being implemented; areas of nonconformity, gaps, and shortcomings; good practices introduced or implemented in similar projects or in the industry; ways to improve implementation of processes to help the team increase productivity; and highlights contributions of the audit in the lessons-learned repository. [Executing]

PMI®, *PMBOK® Guide*, 2013, 247
PMI® *PMP Examination Content Outline,* 2015, Executing, 8, Task 3

18. b. Defect repairs

Change requests are an input to the Perform Integrated Change Control process. They also may include corrective action and preventive action. It should be noted that corrective and preventive actions tend not to affect the project baselines in most cases but do affect performance against the baselines. [Monitoring and Controlling]

PMI®, *PMBOK® Guide*, 2013, 97
PMI® *PMP Examination Content Outline,* 2015, Monitoring and Controlling, 9, Task 2

19. d. Identified in the project management plan

Usually, the project manager or the project sponsor can approve or reject a documented change request. At times, a Change Control Board (CCB) is used, which later may require customer or sponsor approval. Regardless, the responsible person is identified in the project management plan or by organizational procedures. [Monitoring and Controlling]

PMI®, *PMBOK® Guide*, 2013, 96
PMI® *PMP Examination Content Outline,* 2015, Monitoring and Controlling, 9, Task 1

20. d. Change control meetings

    Meetings, referred to as change control meetings, are a tool and technique in Integrated Change Control. Often, a project will set up a Change Control Board, which has the responsibility for meeting and reviewing the change requests, and approving, rejecting, or other disposition of the changes. Decisions of the board are documented and agreed upon by appropriate stakeholders in the change management plan. The CCB decisions are documented to stakeholders for information and follow-up actions. [Monitoring and Controlling]

    PMI®, *PMBOK® Guide*, 2013, 99
    PMI® *PMP Examination Content Outline,* 2015, Monitoring and Controlling, 9, Task 2

21. b. Adds business value as it links to business and project objectives

    The requirements traceability matrix is a grid that links requirements to their origin and traces them throughout the life cycle. It is an output in Collect Requirements This approach helps to ensure that each requirement adds value as it links to the business and project objectives. It also tracks requirements during the life cycle to help ensure that the requirements listed in the requirements document are delivered at the end of the project. It provides a structure to manage changes to product scope. See Figure 5-6 in the *PMBOK® Guide*, 2013 for an example. [Planning]

    PMI®, *PMBOK® Guide*, 2013, 118–119
    PMI® *PMP Examination Content Outline,* 2015, Planning, 6, Task 3

22. b. Enterprise environmental factors

    They are an input to the Direct and Manage Project Work process. Other examples are organizational, cultural, and customer culture, infrastructure, personnel administration, and the PMIS. An example of stakeholder risk tolerances is allowable cost overrun percentages. [Executing]

    PMI®, *PMBOK® Guide*, 2013, 82
    PMI® *PMP Examination Content Outline,* 2015, Executing, 8, Task 7

23. b. Configuration management system

    The formal configuration management system is an important part of Perform Integrated Change Control and focuses on the specifications for deliverables and processes. Its activities involve configuration identification, configuration status accounting, and configuration verification. [Monitoring and Controlling]

    PMI®, *PMBOK® Guide*, 2013, 96–97
    PMI® *PMP Examination Content Outline,* 2015, Monitoring and Controlling, 9, Task 2

24. a. Root cause analysis

Determining the root cause of the problem means to determine the origin of the problem. What may appear to be the problem on the surface is often revealed, after further analysis, not to be the real cause of the problem. Process analysis includes root cause analysis used to identify as problem, discover the underlying causes that lead to it and develop preventive actions. Process analysis is a tool and technique in Perform Quality Assurance. [Executing]

PMI®, *PMBOK® Guide*, 2013, 247
PMI® *PMP Examination Content Outline,* 2015, Executing, 8, Task 3

25. d. Validate Scope typically precedes Control Quality

Validate Scope focuses on accepting project deliverables, and to be accepted, they must meet the quality requirements. Control Quality is one way to ensure the correctness of the deliverables and meeting the quality specifications for the deliverables, which is why Control Quality typically is done before Validate Scope. Further, the verified deliverables obtained from Control Quality have been reviewed with the customer or sponsor to ensure they are completely satisfied and have received formal acceptance from the customer or sponsor. In the Validate Scope process, verified deliverables are a tool and technique, used to ensure project deliverables are completed and checked for correctness through Control Quality. [Monitoring and Controlling]

PMI®, *PMBOK® Guide*, 2013, 134
PMI® *PMP Examination Content Outline,* 2015, Monitoring and Controlling, 9, Task 3

26. c. Making sure all the activities necessary to satisfy exit criteria for a phase or the entire project are followed

Administrative activities are necessary to close the project or a phase. To do so, the project manager needs to engage the proper stakeholders in the process. The first activity that must be done to close a project or a phase is the answer to the question. The second activity involves transferring the products, services, or results to the next phase or productions or operations—basically transitioning the product, services, or results. The final step is to collect or close project records, audit project success or failure, gather lessons learned, and archive information for future use, hopefully with a knowledge transfer system. [Closing]

PMI®, *PMBOK® Guide*, 2013, 101
PMI® *PMP Examination Content Outline,* 2015, Closing, 10, Task 7

27. c. Identifying the key issues and any suitable alternatives

    Identifying the key issues and suitable alternatives are the purpose of prioritization matrixes, a quality management and control tool used as a tool and technique in Perform Quality Assurance. In a prioritization matrix, these issues and alternatives to be prioritized are considered as a set of decisions for implementation. Then, criteria are prioritized and weighted before they are applied to all possible alternatives to obtain a mathematical score to rate the options. In this question, such an approach would provide a way to better focus attention on areas in need of improvement. [Executing]

    PMI®, *PMBOK® Guide*, 2013, 246
    PMI® *PMP Examination Content Outline,* 2015, Executing, 8, Task 3

28. a. Identifies project assumptions

    The project scope baseline is an input to identifying risks. Project assumptions, which should be enumerated in the project scope statement, are areas of uncertainty, and therefore, potential causes of project risk. The WBS is also part of the scope baseline, and it is a critical source to consider in identifying risks as it facilitates an understanding of risks at the micro and macro levels. Risks can be identified and then tracked at the summary, control account, and/or work package levels. [Planning]

    PMI®, *PMBOK® Guide*, 2013, 322
    PMI® *PMP Examination Content Outline,* 2015, Planning, 6, Task 10

29. c. Prototypes

    Prototypes are used to obtain early feedback on requirements by providing a working model of the expected product before it is built. Stakeholders then can experiment with this model rather than discussing abstract representations of requirements. This approach supports progressive elaboration, because it is used in iterative cycles of mock-up creation, user experimentation, feedback generation, and prototype revision. When enough feedback cycles have been completed, it then is time to move to design or build as the next phase. Prototypes are a tool and technique in Collect Requirements. [Planning]

    PMI®, *PMBOK® Guide*, 2013, 116
    PMI® *PMP Examination Content Outline,* 2015, Planning, 6, Task 2

30. a. Control Quality and Plan Quality Management

Data from Plan Quality Management and Control Quality are used in Perform Quality Assurance. The Perform Quality Assurance process also uses affinity diagrams, process decision program charts, interrelationship digraphs, tree diagrams, prioritization matrices, activity network diagrams, matrix diagrams, quality audits, and process analysis. [Executing]

PMI®, *PMBOK® Guide*, 2013, 245–247
PMI® *PMP Examination Content Outline,* 2015, Executing, 8, Task 3

31. d. Continuously monitor the project

The Monitor and Control Project Work process is performed throughout the project and includes collecting, measuring, and disseminating performance information and assessing measurements and trends to effect process improvement. Continuous monitoring is important because it provides insight into the project's health, highlighting areas requiring special attention. [Monitoring and Controlling]

PMI®, *PMBOK® Guide*, 2013, 88
PMI® *PMP Examination Content Outline,* 2015, Monitoring and Controlling, 9, Task 1

32. d. Determine whether the project should continue to the next phase

The review at the end of a project phase is called a phase-end review. It also may be called a stage gate, milestone, phase review, or kill point. The purpose of this review is to determine whether the project should continue to the next phase for detecting and correcting errors while they are still manageable and for ensuring that the project remains focused on the business need it was undertaken to address. It is important in the Develop Project Charter phase as the charter formally authorizes the project, providing the project manager with the authority to apply resources to the project. The charter then serves as an excellent review point before it is approved to determine if the project should move into planning. Clear descriptions of the project objectives may be developed including why a specific project is the best alternative to satisfy requirements. An example of a single process, showing among other things initiating leading to planning, is found in Figure 2-10 in the *PMBOK® Guide*, 2013. [Initiating]

PMI®, *PMBOK® Guide*, 2013, 41–42, 55, 71, 549
PMI® *PMP Examination Content Outline,* 2015, Initiating, 5, Task 5

33. c. Celebrating

    During the adjourning stage of team development, the emphasis is on tasks and relationships that promote closure and celebration. There is recognition and satisfaction as the team is moving on and separation. Management skills involve evaluating, reviewing, and improving, while leadership qualities are celebrating and bringing closure. Project staff members are released as deliverables are completed or as part of closure; however these phases of team development (others are forming, storming, norming, and performing) are part of team-building techniques, a tool and technique in the Develop Project Team process. [Executing]

    Verma, V.K., *Managing the Project Team*, 1997, 40
    PMI®, *PMBOK® Guide*, 2013, 276
    PMI® *PMP Examination Content Outline,* 2015, Executing, 8, Task 2

34. b. At +$300, the situation is favorable, as physical progress is being accomplished ahead of your plan.

    Schedule variance is calculated as EV – PV, or $1,500 – $1,200 = +$300. Because the SV is positive, physical progress is being accomplished at a faster rate than planned. It is a useful metric because it can indicate when a project is failing behind or is ahead of its baseline schedule and should be used along with critical path analysis. [Monitoring and Controlling]

    PMI®, *PMBOK® Guide*, 2013, 218, 224
    PMI® *PMP Examination Content Outline,* 2015, Monitoring and Controlling, 9, Task 1

35. d. Managing the approved cost baseline and any changes to it

    The Control Costs process involves monitoring the project's status to update the project costs and managing changes to the cost baseline. Its benefit is that is provides the means to recognize variance in order to take corrective action and minimize risks. Therefore, effective management of the approved cost baseline and any changes is imperative. [Monitoring and Controlling]

    PMI®, *PMBOK® Guide*, 2013, 215–216
    PMI® *PMP Examination Content Outline,* 2015, Monitoring and Controlling, 9, Task 1

36. d. Recognize that your original estimates were fundamentally flawed, and your project is in an atypical situation

CPI = EV/AC. It measures the cost budgeted resources. It is considered the most critical earned value management metric since it measures the cost efficiency for the completed work. The CPI is useful for determining project status and provides a basis to estimate project cost and schedule outcomes. A CPI of 0.44 means that for every dollar spent, you are only receiving 44 cents of progress. Therefore, something is not correct with how you planned your project, or your original estimates were fundamentally flawed, and your project is in an atypical situation. You might want to reconsider a formal "replant," taking a new baseline of your project, or both. [Monitoring and Controlling]

PMI®, *PMBOK® Guide*, 2013, 219, 224
PMI® *PMP Examination Content Outline,* 2015, Monitoring and Controlling, 9, Task 1

37. c. Project scope statement

The project scope statement describes in detail the deliverables and what work must be done to prepare them. Ancillary results are also considered deliverables and are included in the project scope statement. They include items such as project management reports and documentation. Deliverables in the project scope statement may be described at a summary level or in a detailed way. It is the output of the Define Scope process. It also describes the project scope, assumptions, and constraints and covers the entire scope for both the project and the product. [Planning]

PMI®, *PMBOK® Guide*, 2013, 123
PMI® *PMP Examination Content Outline,* 2015, Planning, 6, Task 2

38. c. Expert judgment

According to the *PMBOK® Guide*, expert judgment is used in Close Project or Phase to ensure closure is performed to appropriate standards. Expertise is provided by other project managers in the organization, the PMO, and professional and technical associations. [Closing]

PMI®, *PMBOK® Guide*, 2013, 102
PMI® *PMP Examination Content Outline,* 2015, Closing, 10, Task 1

39. a. Project charter

Outputs of the Define Scope process are the project scope statement and project document updates that include updates to the stakeholder register, requirements documentation, and the requirements traceability matrix. [Planning]

PMI®, *PMBOK® Guide*, 2013, 125
PMI® *PMP Examination Content Outline,* 2015, Planning, 6, Task 2

40. a. Informing your stakeholders of approved change requests and their costs

A number of activities are involved in the Control Costs process. The key to its effectiveness is in managing the approved cost baseline and any changes to it. Since change is inevitable on projects, it is a best practice each time there is an approved change to notify stakeholders of the costs to implement it, which tracks to this question as management is interested in a focus on controlling costs. . [Monitoring and Controlling]

PMI®, *PMBOK® Guide*, 2013, 216
PMI® *PMP Examination Content Outline,* 2015, Monitoring and Controlling, 9, Task 1

41. a. Commence as scheduled and stated in the staffing management plan

Training is a tool and technique for the Develop Project Team process. The requirements and schedule for training needs should be stated in the staffing management plan. A training plan can be developed for the project, and this plan can enable team members to develop or enhance competencies, as well as obtain certifications to benefit the project. In the Develop Project Team process, training is a tool and technique to enhance the competencies of the team members. Further, some project team members' skills can be developed as part of the project activities. Training costs can be in the project's budget or supported by the performing organization especially if the skills acquired can be used on future projects. [Executing]

PMI®, *PMBOK® Guide*, 2013, 266, 275
PMI® *PMP Examination Content Outline,* 2015, Executing, 8, Task 2

42. d. The range of acceptable variances will tend to decrease over time

Variance analysis, along with trend analysis, is examples of performance reviews, a tool and technique in Control Costs. The most frequently analyzed measurements are cost and schedule variances. Cost performance measurements are used to assess the magnitude of variation to the original cost baseline. It is necessary to determine the cause and degree of variance relative to the baseline, but over time, the percentage range of acceptable variation decreases as stakeholders see more work is being accomplished. They also become less likely to terminate the project recognizing its sunk costs and the work completed to date [Monitoring and Controlling]

PMI®, *PMBOK® Guide*, 2013, 222
PMI® *PMP Examination Content Outline,* 2015, Monitoring and Controlling, 9, Task 1

43. c. Estimate at completion

EAC is the expected total cost to complete all work expressed as the sum of the actual costs to date and the estimate to complete (ETC) or the expected cost to finish all remaining work. It can be calculated several different ways. To use it effectively, the project team must determine the ETC based on experience to date. It may differ from the budget at completion. [Monitoring and Controlling]

PMI®, *PMBOK® Guide*, 2013, 220, 224
PMI® *PMP Examination Content Outline,* 2015, Monitoring and Controlling, 9, Task 1

44. d. Links the project to the ongoing work of the organization

The project charter not only authorizes a project, it shows how the project is linked to the strategic plan of the organization, as well as to its ongoing work. Among other things, the project charter documents the business need for the project and describes the current understanding of the requirements. [Initiating]

PMI®, *PMBOK® Guide*, 2013, 68
PMI® *PMP Examination Content Outline,* 2015, Initiating, 5, Task 1

45. a. Meeting management

    Facilitation techniques are a tool and technique in Develop Project Management Plan process. Other examples are brainstorming and problem solving. They are used to help teams and individuals achieve agreement to accomplish the project's objectives. [Planning]

    PMI®, *PMBOK® Guide*, 2013, 76
    PMI® *PMP Examination Content Outline,* 2015, Planning, 6, Task 1

46. d. Process measurement data base

    The process measurement data base is an organizational process asset that is used to collect and make available measurement data on processes and products. The other answers are examples of enterprise environmental factors used as inputs to Direct and Manage Project Work. [Executing]

    PMI®, *PMBOK® Guide*, 2013, 83
    PMI® *PMP Examination Content Outline,* 2015, Executing, 8, Task 2

47. c. EAC = AC + Bottom-up ETC

    This formula assumes that all of the remaining work is independent of the burn rate incurred thus far. AC is $2,900 + [$500 + $1,000]. The $500 is from Activity B, and the $1,000 is from Activity C. This bottom-up approach builds on the actual costs and experience incurred for the work completed and requires a new estimate to complete the remaining work. [Monitoring and Controlling]

    PMI®, *PMBOK® Guide*, 2013, 220–221, 224
    PMI® *PMP Examination Content Outline,* 2015, Monitoring and Controlling, 9, Task 1

48. a. Certain deliverables or subprojects will be accomplished far into the future

    Many projects involve deliverables or subprojects that will be accomplished far into the future and cannot be specified in detail at the current time. In these situations, the project management team typically waits until the deliverable or subproject is clarified so that details for that portion of the WBS can be developed. Then a rolling wave planning approach can be used. [Planning]

    PMI®, *PMBOK® Guide*, 2013, 131
    PMI® *PMP Examination Content Outline,* 2015, Planning, 6, Task 1

49. c. Manage Communications process

Lessons learned documentation is an output of the Manage Communications process. It is an element of the organizational process assets updates. It includes the causes of issues, reasons for corrective actions selected, and other types of lessons learned about communications management. [Executing]

PMI®, *PMBOK® Guide*, 2013, 303
PMI® *PMP Examination Content Outline,* 2015, Executing, 8, Task 6

50. a. Influencing the factors that create change to the authorized cost baseline

The Control Costs process is concerned with ensuring that requested changes have been acted upon, managing actual changes if and when they occur, ensuring cost expenditures do not exceed authorized funding, monitoring cost performance, preventing unapproved changes from being included in the reported cost or resource use, informing stakeholders of all approved changes and their costs, and bringing expected cost overruns within acceptable limits. [Monitoring and Controlling]

PMI®, *PMBOK® Guide*, 2013, 216
PMI® *PMP Examination Content Outline,* 2015, Monitoring and Controlling, 9, Task 2

51. a. Respect and trust

Leadership is critical to project management as it focuses on ensuring a group of people are working toward a common goal and enables them to work as a team. It involves getting things done through others. Respect and trust, not fear and submission, are its key elements. Building trust with the team and other stakeholders is a critical component of effective leadership. Trust is associated with cooperation, sharing information and effective problem resolution. Without trust it is difficult to establish the positive relationships necessary among stakeholders engaged in the project. If trust is compromised, relationships deteriorate, people disengage, and collaboration becomes difficult or impossible. The Develop Project Team process notes the importance of developing trust among team members and the necessity of ensuring the team can work independently in a climate of mutual trust, whether it is colocated or virtual. By improving feelings of trust among team members, morale and team work are increased, and there are fewer conflicts. [Executing]

PMI®, *PMBOK® Guide*, 2013, 274, 513, 517–518
PMI® *PMP Examination Content Outline,* 2015, Executing, 8, Task 1

52. b. 100% rule

The WBS is a deliverable-oriented, hierarchical decomposition of work to be done by the project team. Sometimes called the 100% rule, it shows the total of the work at the lowest levels must roll up to the higher levels so that nothing is left out, and no extra work is done. It shows all product and project work including the project management work. [Planning]

PMI®, *PMBOK® Guide*, 2013, 131
PMI® *PMP Examination Content Outline,* 2015, Planning, 6, Task 2

53. b. Use fast tracking

Fast tracking or crashing the schedule for the remaining work to be done are examples of schedule compression techniques to find ways to bring project activities that are behind into alignment with the project management plan. It is a tool and technique in Control Schedule. In fast tracking, activities or phases normally done in sequence are done in parallel for part of their duration. It may result in rework and increased risk. It only works if activities can be overlapped to shorten the project's duration. [Monitoring and Controlling]

PMI®, *PMBOK® Guide*, 2013, 181, 190
PMI® *PMP Examination Content Outline,* 2015, Monitoring and Controlling, 9, Task 1

54. c. $6.42 million

Test: $5M + $960K + $460K = $6.42M; Don't Test: $7M. Decision tree analysis is a tool and technique as part of quantitative risk analysis techniques. Please refer to Figure 11-16 in the *PMBOK® Guide*, 2013 for an example. [Planning]

PMI®, *PMBOK® Guide*, 2013, 339
PMI® *PMP Examination Content Outline,* 2015, Planning, 6, Task 10

55. d. Promoting job satisfaction

Other key values are challenging work; a sense of accomplishment, achievement, and growth; sufficient financial compensation; and other rewards and recognition the team member considers necessary and required. The Develop Project Team process in its discussion of recognition and rewards as a tool and technique points out that people are motivated if they feel they are valued in the organization. It also says most people are motivated if they have an opportunity to grow, accomplish, and apply their professional skills to meet new challenges. One of the key benefits of this process is motivated employees. To do so, the project manager continually motivates team members by providing challenges and opportunities, providing timely feedback, and recognizing and rewarding good performance. It is ongoing throughout the project. [Executing]

PMI®, *PMBOK® Guide*, 2013, 273–274. 277, 515
PMI® *PMP Examination Content Outline,* 2015, Executing, 8, Task 1

56. a. Work performance information

The project's work performance information should document and communicate the CV, SV, CPI, SPI, TCPI, and VAC for the WBS components in particular for specific work packages and control accounts. It is an output of Control Costs. [Monitoring and Controlling]

PMI®, *PMBOK® Guide*, 2013, 225
PMI® *PMP Examination Content Outline,* 2015, Monitoring and Controlling, 9, Task 1

57. d. A calculated EAC value or a bottom-up EAC value is documented and communicated to stakeholders

Cost forecasts are another output of Control Costs, and the EAC is used to show the expected total costs of completing all work expressed as the sum of the actual cost to date and the estimate to complete. [Monitoring and Controlling]

PMI®, *PMBOK® Guide*, 2013, 224–225
PMI® *PMP Examination Content Outline,* 2015, Monitoring and Controlling, 9, Task 1

58. d. Customer request

Projects can be authorized as a result of a market demand, organizational need, customer request, technological advance, legal requirement, ecological impact, or a social need. The new industrial park is an example of a project authorized because of a customer request. [Initiating]

PMI®, *PMBOK® Guide*, 2013, 69
PMI® *PMP Examination Content Outline*, 2015, Initiating, 5, Task 1

59. a. An analytical technique

Trend analysis, along with regression analysis, is an example of an analytical technique used in Close Project or Phase. It examines any changes in the cost, budget, schedule, and scope baselines during the project to determine why these changes were made and their impact. Then, these impacts can be analyzed and incorporated into the project's lessons learned as part of the organizational process assets updates, an output of this process. Throughout the project, regular use of trend analysis is recommended to see if performance on the project is improving or deteriorating. [Closing]

PMI®, *PMBOK® Guide*, 2013, 91–92, 103, 566
PMI® *PMP Examination Content Outline*, 2015, Closing, 10, Task 5

60. d. List of potential responses

The primary outputs from Identify Risks are initial entries into the risk register. It ultimately contains outcomes of other risk management processes as they are conducted. As an output of Identify Risks, the risk register should contain a list of identified risks and a list of potential responses. The potential responses identified at this time then are used as inputs to Plan Risk Responses. [Planning]

PMI®, *PMBOK® Guide*, 2013, 327
PMI® *PMP Examination Content Outline*, 2015, Planning, 6, Task 10

61. b. Technical success

Team performance assessments are an output of Develop Project Team. Team assessment criteria should be identified by relevant people in this process. The performance of a successful team is measured in terms of technical success according to agreed-upon objectives in the project, including quality levels; performance on the project schedule, whether it finished on time; and performance on budget, whether it finished within financial constraints. [Executing]

PMI®, *PMBOK® Guide*, 2013, 278–279
PMI® *PMP Examination Content Outline*, 2015, Executing, 8, Task 2

62. b. Monte Carlo analysis

Simulation is a tool and technique for the Develop Schedule process by which multiple project durations with different sets of activity assumptions are calculated. Monte Carlo analysis is the most commonly used simulation technique. It is also used as a tool and technique in the Perform Quantitative Risk Analysis process. For schedule risk analysis, a schedule network diagram and duration estimates are used. [Planning]

PMI®, *PMBOK® Guide*, 2013, 180, 340, 547
PMI® *PMP Examination Content Outline,* 2015, Planning, 6, Task 4

63. b. Performance measurement baseline

The PMB is an approved, integrated scope-schedule-cost plan for the project work against which project execution is compared in order to measure and manage performance. It includes contingency reserve but not management reserve. It typically integrates scope, schedule, and cost parameters of the project, but it may also include technical and quality parameters. [Monitoring and Controlling]

PMI®, *PMBOK® Guide*, 2013, 302, 549
PMI® *PMP Examination Content Outline,* 2015, Monitoring and Controlling, 9, Task 1

64. d. Resource breakdown structure

Schedule data for the project schedule includes a number of items as it collects the information used to describe and control the schedule. These data are an output of the Develop Schedule process. They include schedule milestones, schedule activities, activity attributes, and assumptions and constraints. They may include resource histograms, alternate schedules, scheduling contingency reserves, cash flow projections, and order and delivery schedules. In Control Schedule, schedule data are also an output as part of updates to project documents. Here new project network diagrams may need to be developed to display remaining deviations and approved modifications to the schedule. The resource breakdown structure is used in Estimate Activity Durations. It is a hierarchical structure of the identified resources by resource category and type. [Planning and Monitoring and Controlling]

PMI®, *PMBOK® Guide*, 2013, 168, 184, 191, and 561
PMI® *PMP Examination Content Outline,* 2015, Planning, 6, Task 4
PMI® *PMP Examination Content Outline,* 2015, Monitoring and Controlling, 8, Task 1

65. a. Root causes of issues

In the Manage Stakeholder Engagement process, an issue log is an output. This log is updated as new issues are identified, and current issues are resolved. In updating organizational process assets, lessons-learned documentation notes the importance of documenting the root cause analysis of issues faced, reasons corrective actions were chosen, and other types of lessons learned about stakeholder management. These lessons learned are documented and distributed and are part of the historical data base for the project and the performing organization. [Executing]

PMI®, *PMBOK® Guide*, 2013, 408-409
PMI® *PMP Examination Content Outline,* 2015, Executing, 8, Task 7

66. c. Work performance information

Work performance information is an output from Control Schedule. Other tools and techniques that are used are resource optimization techniques, modeling techniques, schedule compression, and scheduling tool. [Planning]

PMI®, *PMBOK® Guide*, 2013, 185
PMI® *PMP Examination Content Outline,* 2015, Planning, 6, Task 4

67. c. Update the project management plan

If there are changes to the stakeholder management strategy, as an output of the Control Stakeholder Engagement process, affected sections of the project management plan are updated because of these changes. Examples are the change management, communications management, cost management, human resource management, procurement management, quality management, requirements management, risk management, schedule management, scope management, and stakeholder management plans. [Monitoring and Controlling]

PMI®, *PMBOK® Guide*, 2013, 413–414
PMI® *PMP Examination Content Outline,* 2015, Monitoring and Controlling, 9, Task 1

68. a. Configuration status accounting

Configuration status accounting captures, stores, and accesses the needed configuration information to manage products and product information effectively. It includes a list of approved configuration identification, status of proposed changes to the configuration and the implementation status of approved changes. [Monitoring and Controlling]

PMI®, *PMBOK® Guide*, 2013, 97
PMI® *PMP Examination Content Outline,* 2015, Monitoring and Controlling, 9, Task 2

69. c. Final output of creating the WBS is described in terms of verifiable products, services, or results.

By using decomposition, the upper-level WBS components are subdivided for the work for each of the deliverables or subcomponents into its most fundamental elements, where the WBS components then represent verifiable products, services, or results. Decomposition is a tool and technique in Create WBS. [Planning]

PMI®, *PMBOK® Guide*, 2013, 131
PMI® *PMP Examination Content Outline,* 2015, Planning, 6, Task 2

70. a. Is optimized to show relationships between activities

All of the answers are ways to show project schedule network diagrams and are ways to present the project schedule, an output of the Develop Schedule process. A time-scaled diagram includes a time scale and bars that represent the duration of activities with the logical relationships. It is optimized to show the relationships between activities where any number of activities may appear on the same line in sequence. [Planning]

PMI®, *PMBOK® Guide*, 2013, 182
PMI® *PMP Examination Content Outline,* 2015, Planning, 6, Task 4

71. d. Activity name

Activity attributes are an output of the Define Activity process. The components for each activity evolve over time. In the initial stages of the project, they include the activity ID, WBS ID, and the activity name. Later, when completed, they may include activity codes, description, predecessor and successor activities, logical relationships, leads and lags, resource requirements, imposed dates, constraints, and assumptions. [Planning]

PMI®, *PMBOK® Guide*, 2013, 153
PMI® *PMP Examination Content Outline,* 2015, Planning, 6, Task 4

72. b. An external dependency

Some dependencies are external ones, and they involve a relationship between project activities and non-project activities. In sequencing activities, the project management team must determine which dependencies are external as they are usually outside of the team's control. Other examples of dependencies, a tool and technique in Sequence Activities, are mandatory, discretionary, and internal. [Planning]

PMI®, *PMBOK® Guide,* 2013, 157–158
PMI® *PMP Examination Content Outline,* 2015, Planning, 6, Task 4

73. d. Trend analysis

Trend analysis is used in many control processes in project management. As a tool and technique in the performance review category in Control Schedule, the trend analysis examines the performance of the project over time to determine whether performance is improving or deteriorating. Graphical analysis techniques are valuable in trend analysis to understand performance to date and to compare it to future performance goals in the form of completion dates. [Monitoring and Controlling]

PMI®, *PMBOK® Guide,* 2013, 188
PMI® *PMP Examination Content Outline,* 2015, Monitoring and Controlling, 9, Task 1

74. c. Carrying out the work

The implementation phase (i.e., carrying out the work) is when all interfaces affecting the project must be coordinated and when the product or service of the project is created. In most projects, this phase is also where a large portion of the project budget is spent. It is where the project's deliverables—or any unique and verifiable product, service, result, or capability—required to produce the process, phase, or project are completed. Deliverables are an output of the Direct and Manage Project Work process. [Executing]

PMI®, *PMBOK® Guide,* 2013, 56, 84
PMI® *PMP Examination Content Outline,* 2015, Executing, 8, Task 2

75. b. Accommodating

Open subordination is much like an accommodating or smoothing style of conflict management in which negotiators are more concerned about positive relationships than about substantive outcomes. It can dampen hostility, increase support and cooperation, and foster more interdependent relationships. It concedes one's position to the needs of others to maintain harmony and relationships, emphasizing areas of agreement rather than areas of difference. [Executing]

Verma, V.K., *Human Resource Skills for the Project Manager,* 1996, 157
PMI®, *PMBOK® Guide,* 2013, 283
PMI® *PMP Examination Content Outline,* 2015, Executing, 8, Task 2

76. b. Authorized procurement administrator

The buyer, through its authorized procurement administrator, is responsible for providing the seller with formal written notice of contract completion. The procurement administrator does so when the seller has met all contractual requirements as articulated in the contract. Requirements for formal procurement closure are usually defined in the terms and conditions of the contract, and close procurements are an output of the Close Procurement process. [Closing]

PMI®, *PMBOK® Guide,* 2013, 389
PMI® *PMP Examination Content Outline,* 2015, Closing, 10, Task 1

77. c. Correspondence

Contract terms and conditions often require written documentation of certain aspects of buyer/seller communications. Examples include any warnings of unsatisfactory performance and requests for changes in the contract or clarification. Other organizational process assets to update include payment schedules and requests and seller performance evaluation documentation. [Monitoring and Controlling]

PMI®, *PMBOK® Guide,* 2013, 386
PMI® *PMP Examination Content Outline,* 2015, Monitoring and Controlling, 9, Task 7

78. b. Project charter

Although the project charter cannot stop conflicts from arising, it can provide a framework to help resolve them, because it describes the project manager's authority to apply organizational resources to project activities. The project charter also documents the business needs, justification, objectives, assumptions and constraints, and high-level requirements that the new product, service, or result is to satisfy. [Initiating]

PMI®, *PMBOK® Guide*, 2013, 71–72
PMI® *PMP Examination Content Outline,* 2015, Initiating, 5, Task 5

79. d. Checklists

Checklists are used to verify that the work of the project and its deliverables fulfill a set of requirements. Checklists, an output of Plan Quality Management, are typically component specific and are used to verify that a set of required steps have been performed. Checklists, an input to the Control Quality process, are structured lists that help verify that the work of the project and its deliverables fulfill a set of requirements. [Monitoring and Controlling]

PMI®, *PMBOK® Guide*, 2013, 242 and 250
PMI® *PMP Examination Content Outline,* 2015, Monitoring and Controlling, 9, Task 3

80. c. Zero-sum game analysis

Achieving mutual gain during negotiations means that each party benefits by the decisions made. A zero-sum game is where one side wins at the expense of the other. In negotiating, the strategy is to confer with parties of shared or common interests with a view toward compromise or reaching an agreement. Negotiation is a tool and technique in the Acquire Project Team process as the project management team may need to negotiate staff assignments with functional managers, other project teams in the organization, with external organizations, or all three. One's ability in negotiating and influencing others can assist greatly in acquiring exceptional personnel on the project. [Executing]

PMI®, *PMBOK® Guide*, 2013, 270, 517
PMI® *PMP Examination Content Outline,* 2015, Executing, 8, Task 2

81. d. Control charts

Control charts help to determine whether or not a process is stable or has predictable performance. This function of control charts is achieved through the graphical display of results over time to determine whether differences in the results are created by random variations or are unusual events. In a manufacturing environment, such charts are used to track repetitive actions such as manufactured lots. In a project management environment, they can be used to monitor processes such as cost and schedule variances, number requirements, and errors in project documents. [Monitoring and Controlling]

PMI®, *PMBOK® Guide*, 2013, 238
PMI® *PMP Examination Content Outline,* 2015, Monitoring and Controlling, 9, Task 3

82. d. Multi-criteria decision analysis

As a tool and technique in Acquire Project Team, this approach enables criteria to be developed and used to rate or score potential team members. The criteria are weighted given the team's requirements including items such as availability, cost, experience, ability, knowledge, skills, attitude, and international factors. [Executing]

Verma, V.K., *Human Resource Skills for the Project Manager*, 1996, 216–217
PMI®, *PMBOK® Guide*, 2013, 271–272
PMI® *PMP Examination Content Outline,* 2015, Executing, 8, Task 2

83. d. A defined integrated change control process

If the recommended corrective or preventive actions or a defect repair require a change to any of the project management plan, a change request should be prepared in conformance with the defined Perform Integrated Change Control process, an output of Control Quality. [Monitoring and Controlling]

PMI®, *PMBOK® Guide*, 2013, 253
PMI® *PMP Examination Content Outline,* 2015, Monitoring and Controlling, 9, Task 3

84. a. Help anticipate how problems occur

Flowcharts, one of the seven basic quality tools and a tool and technique in Control Quality, depict the interrelationship of a system's components and show the relationships among process steps. They are often referred to as process maps as they display the sequence of steps and the branching possibilities for a process that transforms one or more inputs into one or more outputs. Flowcharts show activities, decision points, branching loops, parallel paths, and the order of processing. As such, they aid the team in anticipating where quality problems might occur, which helps in developing approaches for dealing with these potential problems. [Monitoring and Controlling]

PMI®, *PMBOK® Guide,* 2013, 236
PMI® *PMP Examination Content Outline,* 2015, Monitoring and Controlling, 9, Task 3

85. a. Escalation procedures for handling issues

Issues occur on projects, and an escalation process to handle them is a best practice within the team and the performing organization. It is also an organizational process asset. The other answers are examples of enterprise environmental factors to consider. [Executing]

PMI®, *PMBOK® Guide,* 2013, 260
PMI® *PMP Examination Content Outline,* 2015, Executing, 8, Task 2

86. c. Adjusting leads and lags

Corrective action is anything that brings expected future schedule performance in line with the project plan. Adjusting leads and lags is one of many tools available to identify the cause of variation in the Control Schedule process. [Monitoring and Controlling]

PMI®, *PMBOK® Guide,* 2013, 190
PMI® *PMP Examination Content Outline,* 2015, Monitoring and Controlling, 9, Task 1

87. d. Defect repair

During the Direct and Manage Project Work process, work performance data are collected, actioned, and communicated. This process requires review of the impact of project changes and the implementation of proposed changes, one of which is defect repair. It is an intentional activity to modify a nonconforming product or project component. [Executing]

PMI®, *PMBOK® Guide,* 2013, 81
PMI® *PMP Examination Content Outline,* 2015, Executing, 8, Task 1

88. c. The person who formally authorizes the project

The charter is issued by the project initiator or sponsor who formally authorizes the existence of a project and provides the project manager with the authority to apply organizational resources to project activities. [Initiating]

PMI®, *PMBOK® Guide*, 2013, 71
PMI® *PMP Examination Content Outline,* 2015, Initiating, 5, Task 6

89. a. Smoothing

Smoothing or accommodating emphasizes areas of agreement while avoiding points of disagreement. It concedes one's position to the needs of others to maintain harmony and disagreements. It tends to keep peace only in the short term. [Executing]

Adams, J.R., et al., *Principles of Project Management,* 1997, 181–189
Verma, V.K., *Human Resource Skills for the Project Manager,* 1996, 118
PMBOK® Guide, 2013, 283
PMI® *PMP Examination Content Outline,* 2015, Executing, 8, Task 2

90. c. Does not require 100% inspection of the components to achieve a satisfactory inference of the population

The application of the statistical concept of probability has proven, over many years in many applications, that an entire population of products need not be inspected, if the sample selected conforms to a normal distribution of possible outcomes (the "bell" curve). Sample frequency and sizes should be determined as the quality management plan is prepared in order that the cost of quality includes the number of tests and expected scrap. It is a tool and technique in Control Quality. [Monitoring and Controlling]

PMI®, *PMBOK® Guide*, 2013, 240 and 252
PMI® *PMP Examination Content Outline,* 2015, Monitoring and Controlling, 9, Task 3

91. c. Use a control chart

A control chart is one of the seven basic tools of quality control that determines whether or not a process is stable or has predictable performance. It also illustrates how a process behaves over time. When a process is within acceptable limits, it need not be adjusted; when it is outside acceptable limits, an analysis should be conducted to determine the reasons why. There may be penalties with exceeding the specification limits. It is an input to the Control Quality process. [Monitoring and Controlling]

PMI®, *PMBOK® Guide*, 2013, 238, 252
PMI® *PMP Examination Content Outline,* 2015, Monitoring and Controlling, 9, Task 3

92. b. You need to reduce activity costs as much as possible

This question shows an opportunity has been recognized, which means the risk register is being used as it is an input to the Estimate Costs process. It should be reviewed to consider risk response costs as risks, whether threats or opportunities, have an impact on both activity and overall project costs. In this example, it is a potential opportunity that can benefit the business either by reducing directly activity costs or accelerating the schedule. [Planning]

PMI®, *PMBOK® Guide*, 2013, 203
PMI® *PMP Examination Content Outline,* 2015, Planning, 6, Task 3

93. a. Cost management plan

The management and control of costs focuses on variances. Certain variances are acceptable, and others, usually those falling outside a particular range, are unacceptable. The actions taken by the project manager for all variances are described in the cost management plan. These thresholds are typically expressed as percentage deviations from the baseline plan. These variances are used to monitor cost performance, [Planning]

PMI®, *PMBOK® Guide*, 2013, 199
PMI® *PMP Examination Content Outline,* 2015, Planning, 6, Task 3

94. c. Norming; high supportive and low directive approach

There are four stages of team development: forming, storming, norming, and performing in the original team development model developed by Tuckman; later adjourning was added as a fifth stage. Different leadership styles in terms of the amount of required supportive and directive behavior are appropriate when a team is in a certain development stage. At the norming stage, the third stage in team development, leaders provide high support and low direction. The team members begin to work together effectively and adjust their work habits and behavior to support the team. The team members trust one another. [Executing]

Verma, V.K., *Human Resource Skills for the Project Manager*, 1996, 227
PMI®, *PMBOK® Guide*, 2013, 276
PMI® *PMP Examination Content Outline,* 2015, Executing, 8, Task 2

95. a. The procurement administrator is reassigned

The Close Procurements process looks at the administration of the contract and not the people responsible or involved with the contract. The key benefit of this process is that it documents agreements and related documentation for future reference. The other answers are examples of administrative activities. [Closing]

PMI®, *PMBOK® Guide*, 2013, 386–387
PMI® *PMP Examination Content Outline,* 2015, Closing, 10, Task 6

96. a. Histogram

A histogram is one of the seven basic quality tools used in the Control Quality process. In a histogram, or a special form of bar chart, each column represents an attribute or characteristic of a problem or situation. The height of each column represents the relative frequency of the characteristic. It describes the central tendency, dispersion, or shape of a statistical distribution. Unlike a control chart, it does not consider the influence of time on the variation that exists within the distribution. [Monitoring and Controlling]

PMI®, *PMBOK® Guide*, 2013, 238, 252
PMI® *PMP Examination Content Outline,* 2015, Monitoring and Controlling, 9, Task 3

97. a. Estimate Costs

The Estimate Activity Resources process involves estimating the type and quantities of material, people, equipment, or supplies needed to perform each activity, which allows more accurate cost and activity durations. This means close coordination with the Estimate Costs process is needed. [Planning]

PMI®, *PMBOK® Guide,* 2013, 160, 162
PMI® *PMP Examination Content Outline,* 2015, Planning, 6, Task 4

98. d. Engaging stakeholders at certain stages

While the other answers are good practices, this answer is part of the Manage Stakeholder Engagement process. It shows the necessity of engaging stakeholders at appropriate project stages to obtain or confirm their continual commitment to the project. If a key stakeholder is missing meetings, it is critical for the project manager to meet with this stakeholder and find out his or her concerns and actively listen. If the stakeholder has concerns, then the project manager should address them and see if he or she can regain the stakeholder's support. [Executing]

PMI®, *PMBOK® Guide,* 2013, 405
PMI® *PMP Examination Content Outline,* 2015, Executing, 8, Task 7

99. c. Use subject matter experts

In the Control Stakeholder Engagement process, expert judgment is a tool and technique. They are used to ensure a comprehensive identification and list of new stakeholders and to reassess current stakeholders. Answer A is something the expert could do but the word 'personally' notes it is something you are doing as the project manager, and it is one of your areas of responsibilities. Answer B is eliminated, as a stakeholder management plan is prepared. Answer D is to be done by the project manager and his or her team, as noted by the word 'your' in the answer, and while the expert reassesses current stakeholders, it goes further than only the power/interest grid. [Monitoring and Controlling]

PMI®, *PMBOK® Guide,* 2013, 412
PMI® *PMP Examination Content Outline,* 2015, Monitoring and Controlling, 9, Task 1

100. d. The buffer needed and the buffer remaining

Critical chain is an approach in scheduling in which the project team can place buffers on any project schedule path to account for limited resources and project uncertainties. During a performance review, comparing the amount of buffer remaining to the amount of buffer needed to protect the delivery date can help to determine schedule status. The difference between the buffer needed and the buffer remaining can determine whether corrective action is appropriate. The critical chain method is part of performance reviews, a tool and technique in Control Schedule. [Monitoring and Controlling]

PMI®, *PMBOK® Guide*, 2013, 178, 189
PMI® *PMP Examination Content Outline,* 2015, Monitoring and Controlling, 9, Task 1

101. b. The resource pool can be limited to those people who are knowledgeable about the project

Resource calendars are an input to the Estimate Activity Resource process and to the Estimate Activity Durations process. They are used to estimate resource use. Early in a project, the resource pool might include people at different levels of expertise in large numbers, but as the project progresses, the resource pool then can be limited to those people who are knowledgeable about the project because of their work on it. [Planning]

PMI®, *PMBOK® Guide*, 2013, 163, 168
PMI® *PMP Examination Content Outline,* 2015, Planning, 6, Task 4

102. d. Bottom-up estimating

When an activity cannot be estimated with a reasonable degree of confidence, the work then needs to be decomposed into more detail. The resource needs are estimated. The estimates then are aggregated into a total quantity for each of the activity's resources through a bottom-up approach, a tool and technique in Estimate Activity Resources. These activities may or may not have dependencies between them. However, when dependencies exist, this pattern of use of resources then is documented in the estimated requirements for each activity. Answer a, or the RBS, an output of this process, refers to a hierarchical representation of resources by category and type. Answer b, published estimating data, is another tool and technique in the Estimate Activity Resource process but is used when organizations publish updated production rates and unit costs for resources for different countries and geographical locations. Answer c, also a tool and technique in this process, involves different methods to accomplish schedule activities. [Planning]

PMI®, *PMBOK® Guide*, 2013, 164
PMI® *PMP Examination Content Outline,* 2015, Planning, 6, Task 4

103. a. Review work performance information

The calculated SV and SPI time performance indicators for WBS components, in particular the work packages and control accounts, are documented and communicated to stakeholders as an output of the Control Schedule process. [Monitoring and Controlling]

PMI®, *PMBOK® Guide*, 2013, 190
PMI® *PMP Examination Content Outline,* 2015, Monitoring and Controlling, 9, Task 1

104. d. Focus on goals to be served

Decision making involves the ability to negotiate and influence the organization and the project management team. This answer is the first guideline; the other answers refer to items in leadership and influencing; all are examples of interpersonal skills of the project manager, a tool and technique in the Manage Project Team process. [Executing]

PMI®, *PMBOK® Guide*, 2013, 283–284
PMI® *PMP Examination Content Outline,* 2015, Executing, 8, Task 7

105. a. A direct way for senior management to formally accept and commit to the project

The three benefits of the Develop Project Charter process are the answer and to have a well-defined project start and create a formal record of the project. The other answers refer to purposes of the charter and its importance to the project. [Initiating]

PMI®, *PMBOK® Guide*, 2013, 66–67
PMI® *PMP Examination Content Outline*, 2015, Initiating, 5, Task 6

106. b. Risk register

The Monitor and Control Risks process includes keeping track of those risks on the watch list. Low-priority risks are inputs to the Monitor and Control Risks process and are documented in the risk register. Other inputs that are part of the risk register include identified risks and risk owners, agreed-upon risk responses, control actions to assess the effectiveness of response plans, specific implementation actions, symptoms and warning signs of risk, residual and secondary risks, and the time and cost contingency reserves. [Monitoring and Controlling]

PMI®, *PMBOK® Guide*, 2013, 350
PMI® *PMP Examination Content Outline*, 2015, Monitoring and Controlling, 9, Task 4

107. a. Project statement of work

The project statement of work describes in a narrative form the products, services, or results that the project will deliver. It documents the relationships between the products, services, and results being created, and the business need the project will address. It references the product scope description as well as the business need and strategic plan. It is an input to the Develop Project Scope process. [Initiating]

PMI®, *PMBOK® Guide*, 2013, 68
PMI® *PMP Examination Content Outline*, 2015, Initiating, 5, Task 2

108. b. User groups

Meetings are a tool and technique in the Close Project or Phase process. They may be face to face, virtual, formal, or informal. They may include project team members and other stakeholders involved or affected by the project. Types of meetings include lessons learned, closeout, user group, or reviews. [Closing]

PMI®, *PMBOK® Guide*, 2013, 103
PMI® *PMP Examination Content Outline*, 2015, Closing, 10, Task 7

109. b. Parametric estimating

Parametric estimating uses statistical relationships between historical data and other variables to calculate an estimate for activity parameters such as cost, budget, and duration. The activity durations then are determined quantitatively by multiplying the quantity of work to be performed by the labor hours per unit of work. This technique can produce higher levels of accuracy depending on the reliability of the data in the model. Analogous estimating, or answer c, uses parameters from previous projects. PERT, answer d, or three-point estimating is used to improve the accuracy of single point activity duration estimating. Answer a does not apply. [Planning]

PMI®, *PMBOK® Guide,* 2013, 169–170
PMI® *PMP Examination Content Outline,* 2015, Planning, 6, Task 4

110. d. Performing

The performing stage of team development is noted by a theme of productivity. Management skills involve consensus building, problem solving, decision making, and rewarding, with leadership shown through management by walking around, stewardship delegation, mentoring, being a futurist, and being a cheerleader/champion. At this time, the team is functioning as a well-organized unit. They are independent and work through issues smoothly and effectively. [Executing]

Verma, V.K., *Managing the Project Team,* 1997, 40
PMI®, *PMBOK® Guide,* 2013, 276
PMI® *PMP Examination Content Outline,* 2015, Executing, 8, Task 2

111. b. Terms and conditions in the contract

The terms and conditions can prescribe specific procedures for the various ways that a contract could be terminated. Early termination can result in mutual agreement between the parties, the default by the seller, or the convenience of the buyer if specified in the contract. The rights and responsibilities of the parties in the event of an early termination are stated in the terminations clause of the contract. The procurement agreement (i.e., an understanding, contract, subcontract, or purchase order) includes terms of agreement, and a contract is a legal relationship subject to remedy in the courts. It contains termination clauses. [Closing]

PMI®, *PMBOK® Guide,* 2013, 377–378, 387
PMI® *PMP Examination Content Outline,* 2015, Closing, 10, Task 3

112. d. Involve stakeholders

In the Close Process or Phase process, procedures are established and documented if a project is terminated before completion. To successfully do so, the project manager, or the buyer as in this question, needs to engage all the proper stakeholders in this process. It is not time to reassign staff, as in answer a, as they should be involved in the administrative activities and in a lessons-learned meeting. In answer b, the procurement department should not have the responsibility to complete all the closing activities as they are the project manager's responsibility. In answer c, the seller requested the termination, and the buyer agreed. [Closing]

PMI®, *PMBOK® Guide*, 2013, 101
PMI® *PMP Examination Content Outline,* 2015, Closing, 10, Task 7

113. b. When the project is in the closing phase

The closing process is when all activities across all process groups are complete. It verifies the project then is read for closure. While a variety of activities are performed, one is to perform team member assessments and release resources. While it may be done in conjunction with the functional manager, answer d, it would not be the case if a projectized structure was used. [Closing]

PMI®, *PMBOK® Guide*, 2013, 58
PMI® *PMP Examination Content Outline,* 2015, Closing, 10, Task 1

114. c. Project charter and stakeholder register

The project charter signifies official sanction by top management and starts the planning, or development, phase. This document formally recognizes the existence of the project and provides the project manager with the authority to apply organizational resources to project activities. The stakeholder register is an output of Identify Stakeholders and also is the other output during the initiating processes. [Initiating]

PMI®, *PMBOK® Guide*, 2013, 71–72, 398
PMI® *PMP Examination Content Outline,* 2015, Initiating, 5, Tasks 3 and 5

115. b. Issue logs

Issues occur on projects, and they should be documented in an issue log. This log is maintained and updated; each issue should be closed or if it no longer is an issue to the project, it should be noted on the log. It then is a project document to update along with requirements documentation, the risk register, and the stakeholder register. Project baselines and the process management plan are examples of items to update in the project management plan, another output of this process, and key performance indicators are examples of performance data collected as part of work performance data, another output of this process. [Executing]

PMI®, *PMBOK® Guide*, 2013, 86
PMI® *PMP Examination Content Outline,* 2015, Executing, 8, Task 5

116. a. A kickoff meeting is recommended

Even if team members already know one another, a kickoff meeting should still be held because the meeting always includes more than meeting team members. Specific expectations for the project can be discussed as well as other important administrative details. Team members can agree upon ground rules as it also gives people an opportunity to express their commitment to the project's objectives. Although it is not mentioned *per se* in the *PMBOK® Guide*, 2013, it was mentioned in earlier editions and is noted in the revised examination content outline. In the Plan Human Resource Management process, meetings are a tool and technique often used to reach consensus among the team on the human resource management plan. [Planning]

Verma, V.K., *Managing the Project Team,* 1997, 135
Meredith, J.R. and Mantel, Jr., S.J., *Project Management: A Managerial Approach*, 2012, 224–225
PMBOK® Guide, 2013, 264
PMI® *PMP Examination Content Outline,* 2015, Planning, 7, Task 12

117. a. Consider every meeting a team meeting, not the project manager's meeting

Team building should be an important a part of every project activity as possible. Given that there are many meetings on projects, each team member should be made to feel that it is his or her meeting and not just the project manager's meeting. This will foster greater contribution by each team member. Team building can be part of any meeting even as a five-minute agenda item as the team-building objective is to help individuals work together effectively as a team. It is ongoing for project success. [Executing]

Verma, V.K., *Managing the Project Team,* 1997, 137
PMI®, *PMBOK® Guide,* 2013, 276
PMI® *PMP Examination Content Outline,* 2015, Executing, 8, Task 1

118. c. Influencing

All are useful skills for project managers. In this situation influencing was necessary as the project manager has little or no direct control over team members as they work in a weak matrix. The ability of the project manager to influence stakeholders in a timely basis is critical to project success. Key influencing skills include being persuasive and clearly articulating points and positions, having active and effective listening skills, being aware of and considering various perspectives in any situation, and gathering relevant and critical information to address issues and reach agreements while maintaining mutual trust. [Monitoring and Controlling]

PMI®, *PMBOK® Guide,* 2013, 284
PMI® *PMP Examination Content Outline,* 2015, Monitoring and Controlling, 9, Task 1

119. a. Periodically review the business case

The key words in the question are "multi-phase project." While all of the answers are best practices because it is a multi-phase project, the business case, an input to the Develop Project Charter, may be periodically reviewed to ensure the project is on track to deliver the business benefits. In the early stages of the project, a periodic review of the business case by the sponsoring organization helps to ensure the project remains in alignment with it. The project manager is responsible for ensuring the project meets the goals of the organization and those of a large set of stakeholders, as defined in the business case. [Initiating]

PMI®, *PMBOK® Guide,* 2013, 69
PMI® *PMP Examination Content Outline,* 2015, Initiating, 5, Task 2

120. a. Conduct a risk audit

The risk audit is a tool and technique in the Control Risks process with two purposes: to assess the effectiveness of risk responses and to evaluate the effectiveness of the risk management process. The project manager is responsible for ensuring these audits are done at the frequency defined in the project risk management plan. They may be part of the regular project review meeting, or there may be separate meetings for the risk audit. [Monitoring and Controlling]

PMI®, *PMBOK® Guide*, 2013, 351
PMI® *PMP Examination Content Outline,* 2015, Monitoring and Controlling, 9, Task 4

121. c. Not be viewed as part of the project life cycle

Projects are efforts that occur within a finite period of time with clearly defined beginnings and ends. Maintenance or operations is ongoing and of an indefinite duration. A maintenance activity, as described in this question, is a separate and distinct undertaking. This question is an example of an operational activity in which repetitive outputs are produced, and resources are assigned to do basically the same set of tasks as in a product life cycle. In this question, it ensures business operations occur efficiently by using the optimum resources needed to meet customer demands. [Initiating]

Frame, J.D., *Managing Projects in Organizations: How to Make the Best Use of Time, Techniques, and People,* 2003, 16–17
PMI®, *PMBOK® Guide*, 2013, 13
PMI® *PMP Examination Content Outline,* 2015, Initiating, 5, Task 7

122. c. Technical performance measurement

Technical performance measurement compares technical accomplishments to date to the project plan's schedule of technical achievement. It requires the definition of objective, quantifiable measures of technical performance to help compare actual results against targets. Deviation, such as less functionality than planned at a key milestone, can help to forecast the degree of success in achieving the project scope. Risk reassessment, answer a, reviews current risks, and the amount and detail of repetition depends on how the project relates to its objectives. Variance analysis, answer b, compares planned results to actual results to see if there are deviations from the baseline plan. Reserve analysis, answer d, is used to compare the amount of contingency reserves remaining to the amount of possible risk remaining to see if they are adequate. All four answers are tools and techniques in Control Risks. [Monitoring and Controlling]

PMI®, *PMBOK® Guide*, 2013, 351–352
PMI® *PMP Examination Content Outline,* 2015, Monitoring and Controlling, 9, Task 4

123. d. Control Risks

A workaround is a form of corrective action, as it is a response to a threat that has occurred for which a prior response had not been planned or was not effective. It is followed in most cases by a change request, which is prepared and submitted to the Perform Integrated Change Control process. These change requests are an output to the Control Risks process. [Monitoring and Controlling]

PMI®, *PMBOK® Guide*, 2013, 353, and 567
PMI® *PMP Examination Content Outline,* 2015, Monitoring and Controlling, 9, Task 4

124. b. Demands

Claims administration is a tool and technique in the Control Procurements process. When the buyer and seller cannot agree, this is also called claims, disputes, or appeals and should be documented, processed, monitored, and managed throughout the contract life cycle. [Monitoring and Controlling]

PMI®, *PMBOK® Guide*, 2013, 384
PMI® *PMP Examination Content Outline,* 2015, Monitoring and Controlling, 9, Task 7

125. a. Procurement performance review

These reviews are a tool and technique of the Control Procurements process, which can include a review of seller-prepared documentation and buyer inspections, as well as quality audits conducted during the seller's execution of the work. They seek to identify performance successes or failures, progress with respect to the contract statement of work, and contract noncompliance. They enable the buyer to quantify the seller's demonstrated ability or inability to perform work. Inspections and audits, answers b and c, are specified in the contract and are required by the buyer to verify compliance in the seller's work processes or deliverables; they also are a tool and technique in this process. A status review meeting, answer d, is an informal session to discuss performance to date. [Monitoring and Controlling]

PMI®, *PMBOK® Guide*, 2013, 383
PMI® *PMP Examination Content Outline,* 2015, Monitoring and Controlling, 9, Task 7

126. a. Add a project buffer

After the critical path is identified using the critical chain method, resource availability is entered, and a resource-constrained schedule results. This schedule may have an altered critical path that is known as the critical chain. The critical chain method adds duration buffers that are non-work schedule activities to manage uncertainty. To protect the target finish date from slippage on the critical chain, a project buffer is placed at the end of the critical chain. [Planning]

PMI®, *PMBOK® Guide*, 2013, 178
PMI® *PMP Examination Content Outline,* 2015, Planning, 6, Task 4

127. a. Codes of conduct

Codes of conduct are an example of an enterprise environmental factor in the area of government standards or industry relations along with quality standards and worker protection standards. The other answers are examples of organizational process assets, another input to the Develop Project Charter process. [Initiating]

PMI®, *PMBOK® Guide*, 2013, 70
PMI® *PMP Examination Content Outline,* 2015, Initiating, 5, Task 1

128. c. Tight matrix

A "tight" matrix refers to team members working in close proximity to one another to enhance their ability to perform as a team. It is used in colocation, a tool and technique in Develop Project Team. It can be temporary or for the entire project. It is a way to post schedules and other items and enhances communications and a sense of community and is a team meeting room or a 'war rom'. On a virtual team, an electronic 'war room' can be established. [Executing]

Verma, V.K., *Managing the Project Team,* 1997, 169
PMI®, *PMBOK® Guide*, 2013, 277
PMI® *PMP Examination Content Outline,* 2015, Executing, 7, Task 1

129. b. Consulting with experts

Analytical techniques, an input to Plan Stakeholder Management, can be used to determine the engagement level of stakeholders from unaware to leading. The engagement level is then documented into a stakeholder assessment matrix to show gaps between the current and desired engagement levels. This matrix then enables actions and communications required to close these gaps identified by the project team using experts. Expert judgment is another tool and technique in this process. [Planning]

PMI®, *PMBOK® Guide*, 2013, 401–402
PMI® *PMP Examination Content Outline,* 2015, Planning, 7, Task 6

130. d. As an input to Develop Project Charter

Historical information is an organizational process access in the Develop Project Charter process. Other organizational process assets are organizational standard processes, policies, and process definitions; templates from other project charters; and the lessons learned data base. [Initiating]

PMI®, *PMBOK® Guide*, 2013, 70
PMI® *PMP Examination Content Outline,* 2015, Initiating, 5, Task 4

131. d. Direct and Manage Project Work

Work performance data containing these examples are an output of Direct and Manage Project Work. They are raw observations and measurements identified as activities are being performed to complete the work of the project. These data often are viewed at the lowest level of detail from which information is derived by other processes. The data then are gathered as the work is done and passed to the controlling processes of the various processes for further analyses. [Executing]

PMI®, *PMBOK® Guide*, 2013, 85
PMI® *PMP Examination Content Outline,* 2015, Executing, 8, Task 6

132. d. Forcing

Forcing, or directing, using power or dominance, implies the use of position power to resolve conflict. It involves imposing one viewpoint at the expense of another. Project managers may use it when time is of the essence, when an issue is vital to the project's well-being, or when they think they are right based on available information. It offers only win-lose situations. Although this approach is appropriate when quick decisions are required or when unpopular issues are an essential part of the project, it puts project managers at risk. [Executing]

Adams, J.R., et al., *Principles of Project Management,* 1997, 181–189
Verma, V.K., *Human Resource Skills for the Project Manager,* 1996, 157
PMI®, *PMBOK® Guide*, 2013, 283
PMI® *PMP Examination Content Outline,* 2015, Executing, 8, Task 2

133. c. Managing remaining buffer durations against the remaining durations of task chains

The purpose of the critical chain method is to modify the project schedule to account for limited resources and project uncertainty. The schedule is built using duration estimates with required dependencies and defined constraints as inputs. Then, the critical path is calculated, and resource availability is entered, which means there is a resource-limited schedule with an altered critical path. Buffers protect the critical chain from slippage, and the size of each buffer accounts for the uncertainty in the duration of the chain of dependent tasks that lead up to the buffer. This method then focuses on managing the remaining buffer durations against the remaining duration of task chains. See Figure 6-19 for an example in the *PMBOK® Guide*, 2013. [Planning]

PMI®, *PMBOK® Guide*, 2013, 178
PMI® *PMP Examination Content Outline,* 2015, Planning, 6, Task 4

134. c. Issue a change request

Change requests are an output of the Manage Project Team process. Staffing changes can affect the rest of the team, even if by choice as in this situation or by an uncontrollable event. When staffing changes affect the project, a change request should be processed through the Perform Integrated Change Control process. While in this question the change involved moving the team member to a different assignment, staffing changes also involve outsourcing the work and replacing team members who leave. [Executing]

PMI®, *PMBOK® Guide*, 2013, 284
PMI® *PMP Examination Content Outline,* 2015, Executing, 8, Task 2

135. d. Identify all potential project stakeholders and their relevant information

Stakeholder analysis is a tool and technique in Identify Stakeholders to gather and analyze information to determine whose interests should be taken into account in the process. The first step is to identify all the potential stakeholders and document their relevant information such as roles, departments, interests, knowledge, expectations, and influence levels. Answer c is the third step in the process. Next the model, or answer b, is determined, and the stakeholder register, or answer a, is the output from this process. [Initiating]

PMI®, *PMBOK® Guide*, 2013, 395–396
PMI® *PMP Examination Content Outline,* 2015, Initiating, 5, Task 3

136. b. Project management plan

Project scope is measured against the project management plan. The project scope statement and scope baseline are subsets of the project management plan. However, the whole plan and all the baselines (cost and schedule) need to be met in addition to the scope. The project management plan is the agreement between the project manager and sponsor and defines what constitutes project completion. [Closing]

PMI®, *PMBOK® Guide*, 2013, 102
PMI® *PMP Examination Content Outline,* 2015, Closing, 10, Task 1

137. c. Determine the formula for calculating the estimate at completion (EAC) for the project

Three recognized earned value rules of performance measurement are to (1) determine the EAC calculation to be used on the project as tracking methodologies are specified and to provide a validity check on the bottom-up EAC, (2) establish the earned value measurement techniques (for example, weighted milestones, fixed formula or percent complete), and (3) define the WBS level at which the measurements of control accounts will be performed. Four methods can be used to calculate the EAC. [Planning]

PMI®, *PMBOK® Guide*, 2013, 199, 224
PMI® *PMP Examination Content Outline,* 2015, Planning, 6, Task 3

138. d. Final product, service, or result transition

All the elements are outputs of the Close Project or Phase process, but the final product, service, or result transition is not part of the organizational process assets. It is an output on its own and involves the product, service, or result that the project was created to produce. [Closing]

PMI®, *PMBOK® Guide*, 2013, 103–104
PMI® *PMP Examination Content Outline,* 2015, Closing, 10, Task 1

139. b. Identify the causes of the conflict

In order to address conflict, people must recognize and acknowledge that conflict exists. The project manager must be able to identify the causes of the conflict and then actively manage it to avoid any negative impacts. Another best practice is to establish common ground or shared goals and then to separate people from the problem. Conflict should be addressed early and usually in private with a direct and collaborative approach. [Executing]

Verma, V.K., *Human Resource Skills for the Project Manager,* 1996, 126
PMI®, *PMBOK® Guide*, 2013, 282–283; 518
PMI® *PMP Examination Content Outline,* 2015, Executing, 8, Task 2

140. d. Determine the total float variance

Performance reviews are a tool and technique used in Control Schedule and include trend analysis, critical path method, critical chain method, and earned value management. In terms of the critical path method, the emphasis is on comparing progress along the critical path to determine schedule status. Variance on the critical path will have a direct impact on the project's end date; evaluating progress of activities or near critical paths can identify schedule risk. Calculating the total float for the network then enables free float also to be calculated, or the amount of time that a schedule activity can be delayed without delaying the early start date of a successor or violating a schedule constraint. After the variance is known, the project team can take corrective action to bring performance in line with the plan. [Monitoring and Controlling]

PMI®, *PMBOK® Guide*, 2013, 176–177, 188–189
PMI® *PMP Examination Content Outline,* 2015, Monitoring and Controlling, 9, Task 1

141. a. Group decision-making techniques

Both processes use inspection. Validate Scope also uses group-decision making techniques to reach a conclusion when the validation is performed by the project team and other stakeholders. Methods to reach a group decision include: unanimity, majority, plurality, and dictatorship. In Control Quality, the other tools and techniques are the seven basic quality tools, statistical sampling, and approved change requests review. [Monitoring and Controlling]

PMI®, *PMBOK® Guide*, 2013, 115, 135, 249
PMI® *PMP Examination Content Outline,* 2015, Monitoring and Controlling, 9, Task 3

142. c. Knowledge

In this situation, the key word is 'customer'. In acquiring resources for your team, you can use multi-criteria decision analysis, a tool and technique in this process. The criteria are used to rate and score team members and can be weighted to show their importance. All of the answers to this question are important, but knowledge is the most critical because it considers whether the team member has relevant knowledge of the customer, focusing on this question. The knowledge criterion also involves similar related projects and nuances of the project environment. Other criteria not shown in the answers include availability, which is not applicable because this is a top project in the company, and one should then be able to acquire the resource; cost is not applicable for the same reason; attitude, which is desirable for being able to work well with the team; and international factors, which is not applicable because the question does not note it is a global project. Experience, ability, and skills are critical, but knowledge is selected because it emphasizes the customer. [Executing]

PMI®, *PMBOK® Guide*, 2013, 271–272
PMI® *PMP Examination Content Outline,* 2015, Executing, 8, Task 2

143. d. Forcing

Forcing or directing is represented by a strong desire to satisfy oneself rather than to satisfy others. It involves imposing one viewpoint at the expense of another and is characterized by a win-lose outcome in which one party overwhelms the other. It pushes one's view at the expense of others, usually through a power position to resolve a conflict. [Executing]

Adams, J.R., et al., *Principles of Project Management*, 1997, 181–189
Verma, V.K., *Human Resource Skills for the Project Manager,* 1996, 118 and 120
PMI®, *PMBOK® Guide*, 2013, 282–283, 518
PMI® *PMP Examination Content Outline,* 2015, Executing, 8, Task 2

144. a. Operations managers

Operations managers are stakeholders on many projects. They deal with producing and managing the products and services of the organization. On many projects, they are responsible after the project is complete and has been formally handed off to them for incorporating the project into normal operations and providing long-term support for the product. During the Identify Stakeholder process, they are people who could impact or be impacted by a decision, activity, or outcome of the project, which is noted by the situation in this question. It is unclear from the question whether it is being done in a manufacturing environment, eliminating answer b; answer c also is eliminated because it is not clear whether outsourcing will be used; and answer d is eliminated because it is not stated whether there are business partners. [Initiating]

PMI®, *PMBOK® Guide*, 2013, 12–14, 33, 393
PMI® *PMP Examination Content Outline*, 2015, Initiating, 5, Task 3

145. a. The choice of media

The choice of media, or the way you deliver the information is as important as what you say. It is important to determine when to communicate in writing versus orally, when to prepare an informal memo or when to use a formal report, and when to communicate face to face or by email, as examples. Other examples involve meeting management techniques; presentation techniques, recognizing the importance of body language in communicating; facilitation techniques to build consensus and overcome obstacles; and listening techniques as active listening and removing barriers that adversely affect communications. [Executing]

PMI®, *PMBOK® Guide*, 2013, 298–299
PMI® *PMP Examination Content Outline*, 2015, Executing, 8, Task 6

146. d. Collaborating

Collaborating or problem solving involves bringing people with opposing views together to reach a solution. When there are too many people involved, it is more difficult to reach a solution, given the multiplicity of perspectives. When the parties involved have mutually exclusive views, another conflict resolution approach is recommended. Confronting is especially useful if there is a culturally diverse team with different perspectives to consider, and by enabling these different perspectives to be discussed, the result may be the best approach to resolve a conflict. [Executing]

Adams, J.R., et al., *Principles of Project Management*, 1997, 181–189
Verma, V.K., *Human Resource Skills for the Project Manager*, 1996, 119
PMI®, *PMBOK® Guide*, 2013, 283, 518
PMI® *PMP Examination Content Outline*, 2015, Executing, 8, Task 2

147. c. Stakeholder register

The stakeholder register is the main output of Identify Stakeholders and contains all the details known at the time related to the stakeholders including identification information, assessment information, and stakeholder classification. The project manager and team should review the stakeholder register regularly since stakeholder identification is ongoing throughout the project, and different stakeholders may have different interests, expectations, or potential influence than those initially identified. [Initiating]

PMI®, *PMBOK® Guide*, 2013, 398
PMI® *PMP Examination Content Outline,* 2015, Initiating, 5, Task 3

148. c. Change requests

Quality improvements to processes and procedures as well as the project and product will result in a change request that will be reviewed and evaluated to allow full consideration of the recommended improvements using the Perform Integrated Change Control process. The key words in this question are "that feeds directly into the Perform Integrated Change Control processes". These change requests are used take corrective or preventive actions or to perform defect repair. The quality management plan and updates to project documents are other outputs of this process. [Executing]

PMI®, *PMBOK® Guide*, 2013, 247–248
PMI® *PMP Examination Content Outline,* 2015, Executing, 8, Task 3

149. d. Basis of estimates

Basis of estimates is an output from the Estimate Costs process. Other tools and techniques are expert judgment, analogous estimating, parametric estimating, bottom-up estimating, project management software, and group decision-making techniques. [Planning]

PMI®, *PMBOK® Guide*, 2013, 204–208
PMI® *PMP Examination Content Outline,* 2015, Planning, 6, Task 3

150. a. Develop Project Charter

In the Develop Project Charter process, an agreement is an input to define initial intentions for the project. They may be known as an understanding, a contract, a subcontract, or a purchase order. In the Develop Project Charter process they may be a service level agreement, letter of agreement, letters of intent, verbal agreement, e-mail, or other written agreements. [Initiating]

PMI®, *PMBOK® Guide*, 2013, 70, 377
PMI® *PMP Examination Content Outline,* 2015, Initiating, 5, Task 5

151. c. Accepted deliverables

Accepted deliverables are an input to the Close Project or Phase. They may include approved product specifications, delivery receipts, and work performance documents. Partial or interim deliverables may be included if it is a phased or canceled project. Other inputs are the project management plan and organizational process assets. The other selections are inputs or tools and techniques for other processes. [Closing]

PMI®, *PMBOK® Guide*, 2013, 102
PMI® *PMP Examination Content Outline,* 2015, Closing, 10, Task 1

152. d. Use variance analysis

Variance analysis is the tool and technique in Control Scope. It focuses on determining the cause and degree of difference between the baseline and actual project performance. The magnitude of the variation from the original scope baseline is assessed through project performance measurements. By using variance analysis, a decision can be made as to whether corrective or preventive action is needed. Answers a and b are variants of inputs to scope control, and answer c is part of work performance information, which is an output from this process. [Monitoring and Controlling]

PMI®, *PMBOK® Guide*, 2013, 139
PMI® *PMP Examination Content Outline,* 2015, Monitoring and Controlling, 9, Task 1

153. c. A high-level requirement

The project charter formally authorizes the existence of the project and provides the project manager with the organizational resources for the project activities. Using agile is an example of a high-level requirement, which is one of the many items included in the charter. [Initiating]

PMI®, *PMBOK® Guide*, 2013, 71–72
PMI® *PMP Examination Content Outline,* 2015, Initiating, 5, Task 5

154. c. Note key stakeholders as parties in the contract

Procurement documents are an input to the Identify Stakeholder process. If the project results from a procurement activity or is based on an established contract, the parties in the contract are key project stakeholders. Others, such as suppliers, are also stakeholders and should be added to the stakeholder list. [Initiating]

PMI®, *PMBOK® Guide*, 2013, 394
PMI® *PMP Examination Content Outline,* 2015, Initiating, 5, Task 3

155. a. Manage them closely

You must manage them closely. High-power/high-interest stakeholders who do not support your project could have a devastating effect on your project. Refer to Figure 13-4 in the *PMBOK® Guide*, 2013 for an example. [Initiating]

PMI®, *PMBOK® Guide*, 2013, 396–397
PMI® *PMP Examination Content Outline,* 2015, Initiating, 5, Task 3

156. b. Maintenance requests

Defect repairs, corrective actions, preventive actions, and updates are types of change requests that occur on a project. Updates involve changes to formally controlled documents and plans to reflect modified or additional ideas to cover. Maintenance requests typically would be outside the scope of the project itself. Change requests can be direct, indirect, and externally or internally initiated, as well as optional and legally or contractually mandated. [Executing]

PMI®, *PMBOK® Guide*, 2013, 85
PMI® *PMP Examination Content Outline,* 2015, Executing, 8, Task 4

157. a. Role

The human resource plan documents roles and responsibilities on the project. A role is the function assumed by or assigned to a person in the project. The court liaison in this question is an example of such a role on a project. Role clarity concerning authority, responsibilities, and boundaries also should be documented to help avoid negative conflicts later in the project. [Planning]

PMI®, *PMBOK® Guide*, 2013, 264
PMI® *PMP Examination Content Outline,* 2015, Planning, 6, Task 5

158. a. Determine who decides the project is a success

Project approval criteria should be documented in the project charter. These criteria include determining what constitutes success, who decides the project is successful, and who signs of on the project. [Initiating]

PMBOK® Guide, 2013, 72
PMI® *PMP Examination Content Outline,* 2015, Initiating, 5, Task 6

159. a. Focus on relationships necessary to ensure success

The project manager has limited time on a project, and his or her time should be used as efficiently and effectively as possible. Therefore, by performing a stakeholder analysis, the project manager can identify the stakeholder relationships that can be leveraged to build coalitions and potential partnerships to enhance project success and to determine relationships that need to be influenced differently at different stages of the project or phase [Initiating]

PMI®, *PMBOK® Guide*, 2013, 395
PMI® *PMP Examination Content Outline,* 2015, Initiating, 5, Task 8

160. c. Have an unambiguous owner for each work package

Regardless of the method used to document each team member's roles and responsibilities, the objective is to ensure that each work package has an unambiguous owner and that all team members have a clear understanding of their roles and responsibilities. For example, an organizational chart or hierarchical format may be used to show high-level roles. The text format provides more detailed information to document responsibilities. The first two answers refer to the resource breakdown structure, and the last answer refers to the organizational breakdown structure. [Planning]

PMI®, *PMBOK® Guide*, 2013, 261–262
PMI® *PMP Examination Content Outline,* 2015, Planning, 6, Task 5

161. c. Process analysis

Process analysis is used as a tool and technique in Perform Quality Assurance. The tools and techniques used during Plan Quality Management are: the seven basic quality tools (cause-and-effect diagrams, flowcharts, checksheets, Pareto diagrams, histograms, control charts, and scatter diagrams), cost of quality, benchmarking, design of experiments, statistical sampling, additional quality planning tools (brainstorming, force field analysis, nominal group technique, and quality management and control tools, [affinity diagrams, process decision program charts, interrelationship diagraphs, tree diagrams, prioritization matrices, activity network diagrams, and matrix diagrams]), and meetings. [Planning]

PMI®, *PMBOK® Guide*, 2013, 235–241 and 245–246
PMI® *PMP Examination Content Outline,* 2015, Planning, 6, Task 8

162. c. Quality policy

The quality policy includes the overall intentions and the direction of the organization regarding quality and as formally expressed by top management. When the performing organization lacks a formal quality policy or when the project involves multiple performing organizations, as in a joint venture, the project management team must develop a quality policy for the project as an input to its quality planning under organizational process assets. The quality management plan then describes how the quality policy will be implemented. [Planning]

PMI®, *PMBOK® Guide*, 2013, 234, 241
PMI® *PMP Examination Content Outline,* 2015, Planning, 6, Task 8

163. b. You may need to compensate the seller for seller preparations and for any completed or accepted work

Early termination of a contract is a special case of procurement closure. The rights and responsibilities of the parties are contained in a termination clause of the contract. Typically such a clause allows the buyer to terminate the whole contract or a portion of it for cause or convenience at any time. In doing so, the buyer may need to compensate the seller for seller's preparations and for any completed and accepted work related to the terminated part of the contract. [Executing]

PMI®, *PMBOK® Guide*, 2013, 378, 387
PMI® *PMP Examination Content Outline,* 2015, Executing, 8, Task 1

164. d. It includes a specific contract management plan.

A contract management plan is not part of an agreement. A procurement management plan is used to describe how the project team will acquire goods and services from outside the organization. An agreement may be called an understanding, a contract, a subcontract, or a purchase order. Regardless of the document's complexity, the contract is a mutually binding legal agreement that obligates the seller to provide the specified products, services, or result, and it obligates the buyer to compensate the seller. It is a legal relationship subject to remedy in the courts. [Executing]

PMI®, *PMBOK® Guide*, 2013, 357, 377–378
PMI® *PMP Examination Content Outline,* 2015, Executing, 8, Task 1

165. d. Project review meetings

Bidders conferences are meetings with prospective sellers prior to the preparation of a bid or proposal to answer questions and clarify issues. They are a tool and technique in the Conduct Procurements process. They are used to ensure the prospective sellers have a clear and common understanding of the procurement requirements, and that no bidders receive preferential treatment. Project review meetings are conducted to assess project performance and status. [Executing]

PMI®, *PMBOK® Guide*, 2013, 375
PMI® *PMP Examination Content Outline,* 2015, Executing, 8, Task 1

166. a. Have a meeting

Meetings are tool and technique in Plan Procurement Management. They are held as research alone may not provide specific information for a procurement strategy without additional information exchange with potential bidders. Through collaborating with potential bidders the organization purchasing the material or service may benefit. Suppliers may benefit to influence a mutually beneficial approach or product. Other tools and techniques are make-or-buy analysis, expert judgment, and market research, [Planning]

PMI®, *PMBOK® Guide*, 2013, 366
PMI® *PMP Examination Content Outline,* 2015, Planning, 6, Task 7

167. c. Project expeditor

A variation of the weak matrix organizational structure, the project expeditor has no formal authority to make or enforce decisions. Nonetheless, the project expeditor must be able to persuade those in authority to maintain the project's visibility so that resources will be allocated as needed to meet the project's schedule, budget, and quality constraints. This approach is considered to be effective in high-technology and research and development environments. However, the expeditor tends to be a staff assistant. He or she cannot personally make decisions or enforce them. In the Plan Human Resource Management process, it is the focus of the organization chart for the project and the role of the project manager. [Planning]

Verma, V.K, *The Human Aspects of Project Management,* 1995, 153–154
PMI®, *PMBOK® Guide*, 2013, 23, 258
PMI® *PMP Examination Content Outline,* 2015, Planning, 6, Task 5

168. c. Can use a flexible leadership style

In the Plan Human Resource Management process, organizational theory is a tool and technique. It notes that different organizational structures have different individual responses, individual performance, and relationship characteristics. It also states that organizational theories may recommend a flexible leadership style that adapts to the changes in a team's maturity level, as in this question, throughout the project life cycle. [Planning]

PMI®, *PMBOK® Guide*, 2013, 263
PMI® *PMP Examination Content Outline,* 2015, Planning, 6, Task 5

169. b. Use experts

Although all four answers are tools and techniques in the Plan Human Resource Management process, experts are the most appropriate answer. Expert judgment is a key tool and technique. It especially applies to this question as it notes expert judgment is used to determine the reporting relationships based on the organization culture. [Planning]

PMI®, *PMBOK® Guide*, 2013, 263
PMI® *PMP Examination Content Outline,* 2015, Planning, 6, Task 5

170. a. Preventive action

The Direct and Manage Project Work process, among other thing, also requires review of the impact of all project changes and the implementation of approved changes. This means preventive action, the answer to this question, which is proactive and intentional. [Executing]

PMI®, *PMBOK® Guide*, 2013, 81
PMI® *PMP Examination Content Outline,* 2015, Executing, 8, Task 4

171. b. Auditing project success or failure and archiving records

The key words in this question are "activities for administrative closure". Administrative closure includes step-by-step methodologies that address: actions and activities necessary to satisfy completion or exit criteria for the phase or the project; actions or activities to transfer the products, services, or results to the next phase or to production or operations; and activities to collect project or phase records, auditing success or failure, gathering lessons learned, and archiving information for future use in the organization. Although these administrative closure activities do address transition of the product, answer d, the acceptance of deliverables is not part of these administrative activities. The same is the case with answer c. While lessons learned are included in these administrative activities, final contractor payments are not done as part of these activities. [Closing]

PMI®, *PMBOK® Guide*, 2013, 101
PMI® *PMP Examination Content Outline,* 2015, Closing, 10, Task 3

172. a. An effective and efficient flow of communications among stakeholders

While the Manage Communications process covers numerous areas, its benefit is to enable this efficient and effective communication flow among stakeholders. To do so, appropriate communications technology is required for the information that is being communicated. Appropriate communications models are needed for many purposes including avoiding and managing any communications barriers and choosing communications methods most appropriate to the project and its stakeholders. These methods ensure that the information that has been created and distributed is received and understood for effective responses. Different methods are available to manage and distribute information, and performance reports are prepared periodically and analyzed to assess project progress, as well as performance and forecast project results. [Executing]

PMI®, *PMBOK® Guide*, 2013, 297, 299–301
PMI® *PMP Examination Content Outline,* 2015, Executing, 8, Task 6

173. a. Adaptive life cycle

The adaptive life cycle is one that is known as change driven or one with agile methods and is set up to respond to high levels of change and ongoing stakeholder involvement. This approach differs from iterative and incremental, even though they are characteristic of the adaptive life cycle, as durations are very rapid, such as a two to four week period, and are fixed in time and cost. This approach is preferred in a rapidly changing environment where requirements and scope are difficult to define in advance and when it is possible to define small incremental improvements, which deliver value to stakeholders. [Executing]

PMI®, *PMBOK® Guide*, 2013, 46
PMI® *PMP Examination Content Outline*, 2015, Executing, 8, Task 6

174. c. Prepare a staff release plan

The staff release plan determines the method and timing of releasing team members. Morale is improved if there are smooth transitions for the staff to upcoming projects, which are already planned. This staff release plan also helps to mitigate human resource risks that may be occurring or arise at the end of a project. Further, the costs of these resources are no longer charged to the project, reducing overall project costs. It is part of the staffing management plan, which is part of the human resource plan. [Planning]

PMI®, *PMBOK® Guide*, 2013, 266
PMI® *PMP Examination Content Outline*, 2015, Planning, 6, Task 5

175. d. Analyze each stakeholder's impact or support and classify them

The second step in the stakeholder analysis process is to analyze the potential impact or support each stakeholder could generate and then to classify the stakeholders to define an approach or strategy. It is important to prioritize the stakeholders to ensure the efficient use of effort to communicate with them and best manage their expectations. Answer b is step one in the analysis process, and answer d is the third and last step. In studying for the PMP, if there are steps in a key area, learn them in case there is a question about when something should be done. [Initiating]

PMI®, *PMBOK® Guide*, 2013, 396
PMI® *PMP Examination Content Outline*, 2015, Initiating, 5, Task 3

176. b. Select communications technology

Although all four listed are tools and techniques in the Manage Communications process (another is communications methods), this project is a virtual one and the choice of technology is an important consideration. It can vary from project to project and throughout the life cycle, but as the project manager the focus is to ensure the choice is important for the information that is communicated. It is the first tool and technique listed. The other tool and technique not listed is to determine the communication models. [Executing]

PMI®, *PMBOK® Guide*, 2013, 300–301
PMI® *PMP Examination Content Outline,* 2015, Executing, 8, Task 6

177. d. Storming

During the storming stage, the team is addressing the work, technical decisions, and the project management approach. However, if team members are not collaborating and open to different ideas and perspectives, the environment becomes counter-productive. It is the second stage in the Tuckman team development model (i.e., forming, storming, norming, performing, and adjourning). [Executing]

PMI®, *PMBOK® Guide*, 2013, 276
PMI® *PMP Examination Content Outline,* 2015, Executing, 8, Task 1

178. b. A power/influence grid

Although a number of classification models are available to help prioritize the key stakeholders, the power/influence grid groups stakeholders based on their level of authority or power and their active involvement or interest in the project. See Figure 13-4 in the *PMBOK® Guide*, 2013 for an example. A study tip is to know these four models and to know which one is the best one to use under which circumstances, in case there is a related question on them on your exam, [Initiating]

PMI®, *PMBOK® Guide*, 2013, 396
PMI® *PMP Examination Content Outline,* 2015, Initiating, 5, Task 3

179. b. Checklists

Checklists are a tool and technique of the Identify Risks process and are developed based on historical information and knowledge, as well as lessons learned from previous projects. The lowest level of the risk breakdown structure can be used. Although checklists are quick and simple to use, it is impossible to build a checklist that completely and exhaustively covers all risks because each project is a unique undertaking. The team also needs to explore other items not on the checklist, and it should be pruned at times to archive or remove items that may no longer apply to the organization's projects. It also should be reviewed during project closure to incorporate new lessons learned and improve it for future projects. The other answers in this question refer to other tools and techniques for Identify Risks. [Planning]

PMI®, *PMBOK® Guide*, 2013, 325
PMI® *PMP Examination Content Outline,* 2015, Planning, 6, Task 10

180. b. An emphasis on the International Organization for Standardization

The basic approach to project quality management in the *PMBOK® Guide* is intended to be compatible with the International Organization for Standardization's (ISO) quality standards. Projects require a quality management plan and need data for quality assurance to demonstrate compliance with the plan. In achieving ISO compatibility, modern quality management seeks to minimize variations and to deliver results that meet defined requirements. These approaches recognize the importance of customer satisfaction, prevention over inspection, continuous improvement, management responsibility, and the cost of quality. [Executing]

PMI®, *PMBOK® Guide*, 2013, 227–229
PMI® *PMP Examination Content Outline,* 2015, Executing, 8, Task 3

181. b. All potential sellers are given equal standing

The bidders conference, contractor conference, vendor conference, or pre-bid conference is a tool and technique in Conduct Procurements. They are meetings between the buyer and prospective sellers prior to submitting a bid or a proposal. Bidders conferences are conducted to ensure all prospective sellers have a clear and common understanding of the requirements. They are not used to prequalify vendors. Thus, all vendors are treated equally. All prospective sellers should hear every question and every answer from the buyer. Fairness is addressed by techniques including collecting questions from bidders or arranging field visits in advance of the conference. Responses to questions can be incorporated into the procurement documents as amendments. [Executing]

PMI®, *PMBOK® Guide,* 2013, 375
PMI® *PMP Examination Content Outline,* 2015, Executing, 8, Task 1

182. b. Building trust

Building trust helps to build the foundation of the relationship and is a critical component in effective team leadership. It is associated with cooperation, information sharing, and effective problem resolution. Without trust, it is difficult to establish positive relationships with the various stakeholders engaged in the project. If trust is compromised, people will disengage, and collaboration becomes more difficult if not impossible. [Executing]

PMI®, *PMBOK® Guide,* 2013, 407, 517
PMI® *PMP Examination Content Outline,* 2015, Executing, 8, Task 7

183. d. Transference

Risk transference is shifting some or all negative impact of a threat and the ownership of the response to the threat to a third party. It does not eliminate the threat posed by an adverse risk. It simply gives another party responsibility for it. It almost always involves paying a risk premium to the party taking on the risk. It is most useful in dealing with financial risk exposure. Transference tools are diverse and include using insurance, performance bonds, warranties, and guarantees. Contracts or agreements also can be used to transfer liability for specified risks in case the buyer has capabilities the seller does not possess. It may be prudent to transfer some work and its concurrent risk back to the buyer. A cost-plus contract can be used to transfer the cost risk to the buyer, while a fixed-price contract is preferred to transfer risk to the seller. It is one of four strategies to use in response to negative risks or threats; the other three are avoidance, mitigation, and acceptance. [Planning]

PMI®, *PMBOK® Guide,* 2013, 344
PMI® *PMP Examination Content Outline,* 2015, Planning, 6, Task 10

184. a. It documents agreements and related documentation for future reference

A number of administrative activities are part of this process, but its key benefit is stated in the answer. The outputs then are closed procurements and updates to organizational process assets focusing on the procurement file, acceptance of deliverables, and documentation of lessons learned. [Closing]

PMI®, *PMBOK® Guide*, 2013, 386, 389
PMI® *PMP Examination Content Outline,* 2015, Closing, 10, Task 5

185. d. Conduct a procurement audit

The procurement audit attempts to identify successes and failures relative to the procurement process especially in terms of the preparation or administration of other procurement contracts on the project or on other projects in the organization. Uncovering and reporting both successes and failures can contribute to the project management knowledge base and improve the quality of project management services. It is a structured review of the procurement process from Plan Procurement Management through Control Procurements. A procurement audit should be conducted as part of the Close Procurements process. It is a tool and technique in this process. [Closing]

PMI®, *PMBOK® Guide*, 2013, 388
PMI® *PMP Examination Content Outline,* 2015, Closing, 10, Task 5

186. c. Risk urgency assessment

Risks that may occur in the near-term need urgent attention. The purpose of the risk urgency assessment is to identify those risks that have a high likelihood of occurring. Indicators of priority may include the probability of detecting a risk, time to affect a risk response, symptoms and warning signs, and the risk rating. Assessing risk urgency can be combined with the risk ranking that is determined from the probability and impact matrix for a final risk severity rating. It is a tool and technique in Perform Qualitative Risk Analysis. [Planning]

PMI®, *PMBOK® Guide*, 2013, 333
PMI® *PMP Examination Content Outline,* 2015, Planning, 6, Task 10

187. d. Delphi technique

When consensus is necessary, the Delphi technique is a frequently used information gathering technique as a tool and technique in Identify Risks. A facilitator first sends out a questionnaire to the experts to solicit ideas. The responses then are summarized and returned to the experts for further comment. Consensus generally is reached after a few such rounds. The Delphi technique helps to reduce bias in the data and the undue influence of one person on the outcome. Other information gathering techniques are brainstorming, interviewing, and root-cause analysis. [Planning]

PMI®, *PMBOK® Guide*, 2013, 324
PMI® *PMP Examination Content Outline,* 2015, Planning, 6, Task 10

188. b. Verifying compliance in the seller's work processes

Inspections and audits are tools and techniques in the Control Procurements process. They are required by the buyer and supported by the seller as specified in the procurement contracts and can be conducted as the project is executed to verify compliance in the seller's work processes or deliverables. If it is part of the contract, some inspection and audit teams can include buyer procurement personnel. [Monitoring and Controlling]

PMI®, *PMBOK® Guide*, 2013, 383
PMI® *PMP Examination Content Outline,* 2015, Monitoring and Controlling, 9, Task 7

189. d. To establish minimum requirements of performance for one or more of the evaluation criteria

Weighting systems are developed and used to help select the best seller as part of the proposal evaluation techniques. By assigning a numerical weight to each evaluation criterion, the buyer can emphasize one area as being more important than another. These proposal evaluation techniques are a tool and technique in the Conduct Procurements process for the other answers to this question. [Executing]

PMI®, *PMBOK® Guide*, 2013, 373, 375
PMI® *PMP Examination Content Outline,* 2015, Executing, 8, Task 1

190. a. Hold meetings

Meetings are a tool and technique in the Control Risks process. Risk management should be an agenda item at periodic status meetings. While the amount of time needed for risk management will vary depending on the identified risks, their priority, and the difficulty of the response, the more often risk management is practiced, the easier it becomes. Frequent discussions about risk make it more likely that risks and opportunities will be identified. [Monitoring and Controlling]

PMI®, *PMBOK® Guide*, 2013, 352
PMI® *PMP Examination Content Outline,* 2015, Monitoring and Controlling, 9, Task 4

191. c. Your fee is generally not subject to appeals

This contract type reimburses the seller for all legitimate costs, but the majority of the fee is earned only based on the satisfaction of broad subjective performance criteria defined and incorporated in the contract. The fee determination is based on subjective determination of seller performance by the buyer; it generally is not subject to appeals. [Planning]

PMI®, *PMBOK® Guide*, 2013, 364
PMI® *PMP Examination Content Outline,* 2015, Planning, 6, Task 7

192. d. A mechanism to communicate and support project decision making is provided

All of the answers are outputs in the Control Risks process, however, work performance information specifically provides a mechanism to communicate and support project decision making. [Monitoring and Controlling]

PMI®, *PMBOK® Guide*, 2013, 353–354
PMI® *PMP Examination Content Outline,* 2015, Monitoring and Controlling, 9, Task 4

193. c. Value delivered by vendors meeting the needs

Other factors to consider include the core capabilities of the organization, the risks associated with meeting the need in a cost-effective way, and capability internally compared with the vendor community. Organizations procuring goods and services analyze the need, identify resources, and then compare procurement strategies if they decide to buy. They evaluate factors that influence the make-or-buy decisions including the answer to this question and the other factors listed in the rationale. The make-or-buy decisions are an input to Control Procurements. [Executing]

PMI®, *PMBOK® Guide*, 2013, 374
PMI® *PMP Examination Content Outline,* 2015, Executing, 8, Task 1

194. a. Open items list

Issues or an open item list are examples of inputs if contract negotiation is an independent process. Outputs are documented decisions. While contract negotiations may need to be a separate process for complex procurements, for simple procurement items, the terms and conditions of the contract can be fixed and nonnegotiable. Procurement negotiations are a tool and technique in Control Procurements. [Executing]

PMI®, *PMBOK® Guide*, 2013, 377
PMI® *PMP Examination Content Outline,* 2015, Executing, 8, Task 1

195. a. Plan Procurement Management

Enterprise environmental factors, which include marketplace conditions that the team needs to be aware of as it develops its plans for purchases and acquisition, are an input to the Plan Procurement Management process. [Planning]

PMI®, *PMBOK® Guide*, 2013, 362
PMI® *PMP Examination Content Outline,* 2015, Planning, 6, Task 7

196. b. Program manager

Organizational strategy provides guidance and direction to project management. Portfolio managers, sponsors, or program managers identify alignment or potential conflicts between organizational strategies and project goals and communicates them to the project manager. If the goals of the project conflict with established organizational strategy, the project manager should document and identify these conflicts as early as possible. At times, the development of an organizational strategy could be the goal of a project rather than a guiding principle. If this is the case, then the project must specify what constitutes an appropriate organizational strategy to sustain the organization. [Initiating]

PMI®, *PMBOK® Guide*, 2013, 15
PMI® *PMP Examination Content Outline,* 2015, Initiating, 5, Task 8

197. c. The contract language describes what is necessary to satisfy the project's needs

All project documents are subject to some type of review and approval. However, a contract is legally binding agreement, as well as some other types of agreements, subject to remedy in the courts, and it is therefore subjected to a more extensive approval process. In all cases, however, the primary focus of this review and approval process is to ensure the contract language describes the products, services, or results that satisfy the identified needs of the project. [Executing]

PMI®, *PMBOK® Guide*, 2013, 357
PMI® *PMP Examination Content Outline,* 2015, Executing, 8, Task 1

198. a. Sender-receiver models

Sender-receiver models incorporate feedback loops to provide opportunities for interaction/participation and remove barriers to communication. They are useful in the Manage Communications process as it goes beyond distributing information and seeks to ensure that the information that is being generated for stakeholders has also been received and understood. It also provides opportunities for stakeholders to request further information. [Executing]

PMI®, *PMBOK® Guide*, 2013, 298
PMI® *PMP Examination Content Outline,* 2015, Executing, 8, Task 6

199. a. Seven percent

Albert Meharabian, a researcher, discovered that words alone account for just seven percent of any message's impact. Vocal tones account for 38 percent of the impact, and facial expressions account for 55 percent of the message. Thus, project managers should use nonverbal ingredients to complement verbal message ingredients whenever possible and should recognize that nonverbal factors generally have more influence on the total impact of a message than verbal factors. Body language is mentioned in the Manage Communications process in terms of presentation techniques, along with visual aids. This lack of nonverbal cues makes project communications in a virtual environment more challenging. [Executing]

Verma, V.K., *Human Resource Skills for the Project Manager,* 1996, 19
PMI®, *PMBOK® Guide*, 2013, 299
PMI® *PMP Examination Content Outline,* 2015, Executing, 8, Task 6

200. d. Technological advance

The business case is an input to the Develop Project Charter process. All four answers are reasons for the business case, but this question shows the new product is needed because of technological advances in mobile phones and smart watches. Other reasons to create a business case are a legal requirement, ecological impacts, or a social need. [Initiating]

PMI®, *PMBOK® Guide*, 2013, 69
PMI® *PMP Examination Content Outline,* 2015, Initiating, 5, Task 5

References

Acker, David D. *Skill in Communication: A Vital Element in Effective Management.* 2nd ed. Ft. Belvoir, Va.: Defense Systems Management College, 1992.

Adams, John R., and Bryan W. Campbell. *Roles and Responsibilities of the Project Manager.* Upper Darby, PA: Project Management Institute, 1982.

Adams John R., et al. *Principles of Project Management.* Newton Square, PA: Project Management Institute, 1997.

The Associated General Contractors of America. *Construction Planning and Scheduling.* Washington, D.C.: The Associated General Contractors of America, 1994.

Bell, Chip R. *Managing as Mentors: Building Partnerships for Learning.* San Francisco: Berrett-Koehler, 1996.

Bicheno, John. *The Quality 50.* Melbourne, Australia: Nestadt Consulting Party, 1994.

Bockrath, Joseph T. *Contracts, Specifications, and Law for Engineers.* 4th ed. New York: McGraw-Hill, 1986.

Brake, Terence, Danielle Medina Walker, and Thomas (Tim) Walker. *Doing Business Internationally: The Guide to Cross-Cultural Success.* 2nd ed. Boston: McGraw-Hill, 2002.

Cable, Dwayne, and John R. Adams. *Organizing for Project Management.* Upper Darby, PA: Project Management Institute, 1982.

Carter, Bruce, Tony Hancock, Jean-Marc Morin, and Ned Robins. *Introducing RISKMAN Methodology: The European Project Risk Management Methodology.* Oxford, England: NCC Blackwell, 1994.

Cavendish, Penny, and Martin D. Martin. *Negotiating and Contracting for Project Management.* Upper Darby, PA: Project Management Institute, 1987.

Cibinic, John, Jr., and Ralph C. Nash, Jr. *Cost-Reimbursement Contracting.* 2nd ed. Washington, D.C.: The George Washington University, National Law Center, Government Contracts Program, 1993.

Cleland, David I., and Lewis R. Ireland. *Project Management: Strategic Design and Implementation.* 5th ed. New York: McGraw-Hill, 2007.

Cohen, Dennis J., and Robert J. Graham. *The Project Manager's MBA: How to Translate Project Decisions into Business Success.* San Francisco: Jossey-Bass, 2001.

Corbin, Arthur L. *Corbin on Contracts.* St. Paul, MN: West Publishing, 1952.

Covey, Stephen R. *The Seven Habits of Highly Effective People: Powerful Lessons in Personal Change.* New York: Free Press, 20013.

Crosby, Philip B. *Quality Is Free: The Art of Making Quality Certain.* New York: McGraw-Hill, 1979.

____. *Quality Without Tears: The Art of Hassle-Free Management.* New York: McGraw-Hill, 1984; reprint, New York: Penguin Books, 1985.

Defense Systems Management College. *Risk Management: Concepts and Guidance.* Ft. Belvoir, VA: Defense Systems Management College, 1989.

DeMarco, Tom, and Timothy Lister. *Peopleware: Productive Projects and Teams.* New York: Dorset House Publishing, 1987.

Dinsmore, Paul C. *Human Factors in Project Management.* Rev. ed. New York: American Management Association, 1990.

Dinsmore, Paul C., and Manuel M. Benitez. "Challenges in Managing International Projects." *AMA Handbook of Project Management,* edited by Paul C. Dinsmore. New York: AMACOM Books, 1993, 463–464.

Dinsmore, Paul C., M. Dean Martin, and Gary T. Huettel. *The Project Manager's Work Environment: Coping with Time and Stress.* Upper Darby, PA: Project Management Institute, 1985.

Dobler, Donald W., and David N. Burt. *Purchasing and Supply Management: Text and Cases.* 6th ed. New York: McGraw-Hill, 1996.

Dreger, J. Brian. *Project Management: Effective Scheduling.* New York: Van Nostrand Reinhold, 1992.

Evans, James R., and William M. Lindsay. *The Management and Control of Quality.* 6th ed. Mason, OH: South-Western, 2005.

Ferraro, Gary P. *The Cultural Dimension of International Business.* 3rd ed. Upper Saddle River, NJ: Prentice Hall, 1998.

Filley, Alan C. *Interpersonal Conflict Resolution.* Glenview, IL: Scott, Foresman, and Co., 1975.

Fisher, Roger, William Ury, and Bruce Patton. *Getting to Yes: Negotiating Agreement Without Giving In.* 2nd ed. New York: Penguin Books, 1991.

Fleming, Quentin W. *Cost/Schedule Control Systems Criteria: The Management Guide to C/SCSC.* Chicago: Probus Publishing, 1988.

____. *Project Procurement Management Contracting, Subcontracting, Teaming.* Tustin, CA: FMC Press, 2003.

Fleming, Quentin W., and Joel M. Koppelman. *Earned Value Project Management.* 4th ed. Newtown Square, PA: Project Management Institute, 2012.

Forsberg, Kevin, Hal Mooz, and Howard Cotterman. *Visualizing Project Management.* New York: John Wiley and Sons, 1996.

Frame, J. Davidson. *Managing Projects in Organizations: How to Make the Best Use of Time, Techniques, and People.* 3rd ed. San Francisco, CA: Jossey-Bass, 2003.

____. *The New Project Management: Tools for an Age of Rapid Change, Corporate Reengineering, and Other Business Realities.* 2nd ed. San Francisco, CA: Jossey-Bass, 2002.

Friedman, Jack P. *Dictionary of Business Terms.* 2nd ed. Hauppauge, NY: Barron's Educational Series, Inc., 1994.

Hirsch, William J. *The Contracts Management Deskbook.* Rev. ed. New York: American Management Association, 1986.

Imai, Masaaki. *Kaizen: The Key to Japan's Competitive Success.* New York: McGraw-Hill, 1986.

International Standards Organization. *Guidance on Project Management.* Geneva, Switzerland. ISO., 2012, ISO 21500:2012.

____. *Quality Management Systems—Fundamentals and Vocabulary.* Geneva, Switzerland. ISO., 2008, ISO 9000:2008.

____. *Standardization and Related Activities—General Vocabulary.* Geneva, Switzerland. ISO., 2004, ISO/IEC 2:2004.

____. *Systems and Software Engineering—System Life Cycle Processes.* Geneva, Switzerland. ISO/IEC., 2008, 15288:2008.

Ireland, Lewis R. *Quality Management for Projects and Programs.* Drexel Hill, PA: Project Management Institute, 1991.

Jentz, Gaylord A., Kenneth W. Clarkson, and Roger LeRoy Miller. *West's Business Law.* 2nd ed. St. Paul, MN: West Publishing, 1984.

Katzenbach, Jon R., and Douglas K. Smith. *The Wisdom of Teams.* New York: HarperBusiness, 1994.

Kerzner, Harold. *Project Management: A Systems Approach to Planning, Scheduling, and Controlling.* 11th ed. Hoboken, NJ: John Wiley & Sons, Inc., 2013.

Kirchof, Nicki S., and John R. Adams. *Conflict Management for Project Managers.* Upper Darby, PA: Project Management Institute, 1989.

Kostner, Jaclyn. *Knights of the Tele-Round Table: Third Millennium Leadership.* New York: Warner Books, 1994.

Levin, Ginger, and Steven Flannes. *Essential People Skills for Project Managers.* Vienna, VA: Management Concepts Inc., 2005.

Levin, Ginger. *Interpersonal Skills for Portfolio, Program, and Project Managers.* Vienna, VA: Management Concepts Inc., 2010.

Lewis, James P. *Project Planning, Scheduling, and Control.* Chicago: Probus Publishing, 1991.

Mansir, Brian E., and Nicholas R. Schacht. *An Introduction to the Continuous Improvement Process: Principles and Practices.* Bethesda, MD.: Logistics Management Institute, 1988.

Martin, Martin D., C. Claude Teagarden, and Charles F. Lambreth. *Contract Administration for the Project Manager.* Upper Darby, PA: Project Management Institute, 1990.

Maslow, Abraham H. *Motivation and Personality.* New York: Harper and Row, 1954.

McGregor, Douglas. *The Human Side of Enterprise.* New York: McGraw-Hill, 1960.

Meredith, Jack R., and Samuel J. Mantel, Jr. *Project Management: A Managerial Approach.* 8th ed. Hoboken, NJ: John Wiley and Sons, 2012.

Pennypacker, James S., ed. *Principles of Project Management: Collected Handbooks from the Project Management Institute.* Sylva, NC: Project Management Institute, 1997.

PMI® (*see* Project Management Institute).

Pritchard, Carl L., *Risk Management: Concepts and Guidance.* 5th ed. Boca Raton, FL: CRC Press, 2015.

Project Management Institute. PMI® Code of Ethics and Professional Conduct. www.pmi.org

____. PMI® *Lexicon of Project Management Terms.* www.pmi.org/lexiconterms. 2012

____. *A Guide to the Project Management Body of Knowledge, (PMBOK® Guide).* 5th ed. Newtown Square, PA: Project Management Institute, 2013.

____. *Organizational Project Management Maturity Model* (OPM3®). 3rd ed. Newtown Square, PA: Project Management Institute, 2013.

____.*PMP Examination Content Outline*, June 2015. www.pmi.org

____. *Practice Standard for Earned Value Management,* 2nd ed. Newtown Square, PA: Project Management Institute, 2011.

____. *Practice Standard for Project Estimating.* Newtown Square, PA: Project Management Institute, 2011.

____. *Practice Standard for Scheduling.* 2nd ed. Newtown Square, PA: Project Management Institute, 2011.

____. *Practice Standard for Work Breakdown Structures*, 2nd ed. Newtown Square, PA: Project Management Institute, 2006.

____. *Project Management Professional (PMP) ® Credential Handbook,* 2015. http://www.pmi.org/

____. *The Standard for Portfolio Management*—3rd Edition. Newtown Square, PA: Project Management Institute, 2013.

____. *The Standard for Program Management*—3rd Edition. Newtown Square, PA: Project Management Institute, 2013.

Rose, Kenneth H. *Project Quality Management: Why, What and How.* Second Edition. Boca Raton, FL: J. Ross Publishing, 2014.

Rosen, Robert, Patricia Digh, Marshall Singer, and Carl Phillips. *Global Literacies: Lesson on Business Leadership and National Cultures.* New York: Simon & Schuster, 2000.

Schmauch, Charles H. *ISO 9000 for Software Developers.* Milwaukee: ASQC Quality Press, 1994.

Soin, Sarv Singh. *Total Quality Control Essentials: Key Elements, Methodologies, and Managing for Success.* New York: McGraw-Hill, 1992.

Stewart, Bennett G. *Best Practice-EVA: The Definitive Guide to Measuring and Maximizing Stakeholder Value.* Hoboken, NJ: John Wiley & Sons, 2013.

Stuckenbruck, Linn C., ed. *The Implementation of Project Management: The Professional's Handbook.* Reading, MA: Addison-Wesley, 1981.

Stuckenbruck, Linn C., and David Marshall. *Team Building for Project Managers.* Upper Darby, PA: Project Management Institute, 1985.

Thamhain, Hans J., and David L. Wilemon. "Conflict Management in Project Life Cycles." *Sloan Management Review* 16, no. 3 (Spring 1975): 31–50.

Verma, Vijay K. *Human Resource Skills for the Project Manager.* Vol. 2 of *The Human Aspects of Project Management.* Upper Darby, PA: Project Management Institute, 1996.

____. *Managing the Project Team.* Vol. 3 of *The Human Aspects of Project Management.* Upper Darby, PA: Project Management Institute, 1997.

____. *Organizing Projects for Success.* Vol. 1 of *The Human Aspects of Project Management.* Upper Darby, PA: Project Management Institute, 1995.

Verzuh, Eric. *The Fast Forward MBA in Project Management.* Hoboken, NJ: John Wiley & Sons, 2005.

Vroom, Victor H. *Work and Motivation.* San Francisco: Jossey-Bass, 1995.

Wideman, R. Max, ed. *Project and Program Risk Management: A Guide to Managing Project Risks and Opportunities.* Preliminary ed. Drexel Hill, PA: Project Management Institute, 1992.

Willborn, Walter, and T. C. Edwin Cheng. *Global Management of Quality Assurance Systems.* New York: McGraw-Hill, 1994.

Youker, Robert. "Communication Styles Instrument: A Team Building Tool." In *The Project Management Institute 1996 Proceedings: Revolutions, Evolutions, Project Solutions,* 27th Annual Seminars and Symposium, Papers Presented October 7–9, 1996, Boston, Mass., 796–799. Upper Darby, PA: Project Management Institute, 1996.